高等院校信息技术规划教材

Android应用开发教程

刘志强 主 编
庄旭菲 张 旭 副主编

清华大学出版社
北京

内 容 简 介

本书从已有 Java 基础的初学者角度出发，以 Android 的应用程序开发为主题，通过通俗易懂的语言，循序渐进、系统地介绍了 Android 平台基础知识以及进行应用程序开发应该掌握的基本技术。全书共分 14 章，内容包括 Android 系统架构、开发环境搭建、Android 开发工具、Android 的界面开发、服务与广播、Android 多媒体、数据存储与数据共享、网络编程、Android 传感器、Android 位置服务与地图应用、综合移动应用项目等。本书的讲述由浅入深，结合了大量的实例，以加深读者对 Android 基础知识和基本应用的理解，帮助读者系统地掌握 Android 应用程序设计的基本技术，为从事基于 Android 的应用软件开发打下坚实的基础。

本书可作为计算机及相关专业基于 Android 平台应用开发的教材，也可供专业技术人员参考。

本书封面贴有清华大学出版社防伪标签，无标签者不得销售。
版权所有，侵权必究。侵权举报电话：010-62782989　13701121933

图书在版编目(CIP)数据

Android 应用开发教程/刘志强主编. --北京：清华大学出版社，2016（2019.1重印）
高等院校信息技术规划教材
ISBN 978-7-302-42817-6

Ⅰ. ①A… Ⅱ. ①刘… Ⅲ. ①移动终端－应用程序－程序设计－高等学校－教材 Ⅳ. ①TN929.53

中国版本图书馆 CIP 数据核字(2016)第 028492 号

责任编辑：张　玥　薛　阳
封面设计：常雪影
责任校对：时翠兰
责任印制：宋　林

出版发行：清华大学出版社
　　　网　　址：http://www.tup.com.cn, http://www.wqbook.com
　　　地　　址：北京清华大学学研大厦 A 座　　邮　编：100084
　　　社 总 机：010-62770175　　　　　　　　　邮　购：010-62786544
　　　投稿与读者服务：010-62776969, c-service@tup.tsinghua.edu.cn
　　　质量反馈：010-62772015, zhiliang@tup.tsinghua.edu.cn
　　　课件下载：http://www.tup.com.cn,010-62795954

印刷者：北京富博印刷有限公司
装订者：北京市密云县京文制本装订厂
经　销：全国新华书店
开　本：185mm×260mm　　　印　张：21.75　　　字　数：503 千字
版　次：2016 年 5 月第 1 版　　　　　　　　　　　印　次：2019 年 1 月第 4 次印刷
印　数：5001～6500
定　价：44.50 元

产品编号：067345-01

前言

由Google公司发起的OHA联盟，在2007年11月推出了开放的Android平台。Android是一个流行、免费、开源的移动终端平台，在与其他移动平台的竞争中，一直保持了高速的增长率，众多开发人员已把Android应用开发列为重点选择。如何迅速地推广和普及Android平台软件开发技术，让越来越多的人参与到Android应用的开发中，是国内整个IT行业都在关注的一个话题。

本书是一本以Android的应用开发为主题的基础教材，面向已经具有Java基础的高等院校的学生、开发人员和移动应用开发爱好者，通过对Android平台基础知识以及应用程序开发的基本技术的讲解，帮助读者迅速地掌握Android应用开发技能，为今后从事基于Android的应用软件开发打下坚实的基础。

本书的主要内容

第1章介绍了Android系统的发展史、系统特性、系统架构及Android的开发环境搭建。

第2章介绍了Android常用的开发调试工具，如何创建、运行Android项目，并对Android项目结构进行分析。

第3章介绍了Android UI常用组件及Android布局方法。

第4章介绍了Activity的生命周期、如何创建与注册Activity、Activity的启动方式。

第5章介绍了Android UI高级组件编程，包括ListView、GridView、Spinner、菜单和对话框。

第6章介绍了Android的图形绘制、音频和视频播放技术、Android的动画设计。

第7章介绍了Android的各种传感器应用。

第8章介绍了Android服务的生命周期、创建及配置方法。

第9章介绍了Android广播机制、注册和收发广播方法。

第10章介绍了Android的偏好设置、文件存储、SQLite数据库存储和ContentProvider。

第11章介绍了Android的网络编程，URLConnection接口和

HttpClient 接口的使用方法。

第 12 章介绍了 Android 电话管理器和短信管理器。

第 13 章介绍了 Android 位置服务与地图应用。

第 14 章介绍了一个"手机监控"综合项目案例。

本书的读者对象

- 高等院校计算机类、电子类、电气类、控制类等专业本科生。
- 学习 Android 应用程序开发的研究生。
- Android 应用程序的开发人员,对移动应用开发有兴趣的爱好者。

作者分工

全书由刘志强主编并编写第 1、2、3、4、8 章,张旭编写第 5、6、7 章,庄旭菲编写第 9、10、11、12、13、14 章。同时刘冬梅、王瑞也参与了本书的编写与校稿。

由于作者水平有限,编写时间仓促,书中难免存在疏漏和不足。敬请读者不吝赐教,对本书给予建议和批评指正,以便我们在改版或再版的时候及时纠正补充。

作　者

2015 年 12 月

目录

第 1 章 Android 概述 ... 1

1.1 Android 的发展史 ... 1
1.1.1 Android 的起源 ... 1
1.1.2 当前的主流移动操作系统 ... 2
1.1.3 Android 的版本 ... 3
1.2 Android 系统的特性 ... 5
1.3 Android 系统架构 ... 7
1.3.1 Android 应用层 ... 8
1.3.2 Android 应用框架层 ... 8
1.3.3 Android 核心库 ... 9
1.3.4 Android 运行时 ... 10
1.3.5 Linux 内核层 ... 11
1.4 Android 环境搭建 ... 11
1.4.1 Java 开发环境安装与配置 ... 11
1.4.2 安装 Android 开发环境 ... 12
习题 1 ... 18

第 2 章 开发工具介绍及项目结构 ... 19

2.1 Android 开发工具介绍 ... 19
2.1.1 adb ... 19
2.1.2 AVD ... 20
2.1.3 DDMS ... 22
2.2 创建第一个 Android 项目 ... 25
2.2.1 如何创建 Android 项目 ... 26
2.2.2 Android 项目结构 ... 30
2.2.3 案例程序分析 ... 31
2.2.4 Android 的应用程序组件 ... 35

习题 2 ……………………………………………………………………………………… 37

第 3 章 Android UI 编程 ……………………………………………………………… 38

3.1 Android 常用 UI 组件 ……………………………………………………………… 38
3.1.1 View 和 ViewGroup ……………………………………………………… 38
3.1.2 Android UI 开发概述 …………………………………………………… 40
3.1.3 文本框与编辑框 ………………………………………………………… 41
3.1.4 按钮与图片视图 ………………………………………………………… 47
3.1.5 案例 ImageView 和 ImageButton ……………………………………… 48
3.1.6 案例 CheckBox、RadioButton 和 ToggleButton ……………………… 52
3.2 Android 常用布局 ………………………………………………………………… 60
3.2.1 线性布局 ………………………………………………………………… 61
3.2.2 相对布局 ………………………………………………………………… 63
3.2.3 表格布局 ………………………………………………………………… 68
3.2.4 帧布局 …………………………………………………………………… 71
3.2.5 网格布局 ………………………………………………………………… 72
3.2.6 绝对布局 ………………………………………………………………… 74

习题 3 ……………………………………………………………………………………… 75

第 4 章 Android 活动简介 …………………………………………………………… 76

4.1 Activity 的创建与注册 …………………………………………………………… 76
4.1.1 Activity 的创建 ………………………………………………………… 76
4.1.2 Activity 的注册 ………………………………………………………… 78
4.2 Activity 的生命周期 ……………………………………………………………… 79
4.3 Activity 的启动 …………………………………………………………………… 81
4.3.1 直接启动 Activity ……………………………………………………… 81
4.3.2 启动 Activity 并传递参数 ……………………………………………… 82
4.3.3 带返回值启动 Activity ………………………………………………… 83
4.4 Activity 的启动模式 ……………………………………………………………… 84
4.5 案例 lifecycle …………………………………………………………………… 87
4.5.1 案例功能描述 …………………………………………………………… 87
4.5.2 案例程序结构 …………………………………………………………… 87
4.5.3 案例的实现步骤和思路 ………………………………………………… 87
4.5.4 案例参考代码 …………………………………………………………… 88
4.5.5 案例运行效果 …………………………………………………………… 97
4.5.6 案例程序分析 …………………………………………………………… 98

习题 4 ……………………………………………………………………………………… 101

第 5 章 Android 高级 UI 编程 ………………………………………………… 102

5.1 Adapter 简介 …………………………………………………………… 102
5.1.1 ArrayAdapter 简介 ………………………………………………… 103
5.1.2 案例 ArrayAdapter ………………………………………………… 103
5.1.3 SimpleAdapter 简介 ……………………………………………… 105
5.1.4 案例 SimpleAdapter ……………………………………………… 106

5.2 ListView 列表控件的功能及使用 ……………………………………… 109
5.2.1 ListView 常用属性 ……………………………………………… 109
5.2.2 案例 ListView 具体使用 ………………………………………… 111
5.2.3 响应单击事件 …………………………………………………… 117

5.3 GridView 网格控件的功能及使用 ……………………………………… 118
5.3.1 GridView 常用属性 ……………………………………………… 118
5.3.2 案例 GridView 具体使用 ………………………………………… 118

5.4 Spinner 的功能及使用 ………………………………………………… 122
5.4.1 案例功能描述 …………………………………………………… 122
5.4.2 案例程序结构 …………………………………………………… 122
5.4.3 案例的实现步骤和思路 ………………………………………… 122
5.4.4 案例参考代码 …………………………………………………… 122
5.4.5 案例运行效果 …………………………………………………… 124

5.5 菜单 Menu ……………………………………………………………… 125
5.5.1 使用 xml 定义 Menu …………………………………………… 125
5.5.2 使用代码定义 Menu …………………………………………… 126
5.5.3 使用菜单 ………………………………………………………… 127

5.6 案例菜单 Menu ………………………………………………………… 127
5.6.1 案例功能描述 …………………………………………………… 127
5.6.2 案例程序结构 …………………………………………………… 127
5.6.3 案例的实现步骤和思路 ………………………………………… 127
5.6.4 案例参考代码 …………………………………………………… 128
5.6.5 案例运行效果 …………………………………………………… 130

5.7 对话框 Dialog ………………………………………………………… 130
5.7.1 简单对话框 ……………………………………………………… 131
5.7.2 多按钮对话框 …………………………………………………… 131
5.7.3 列表对话框 ……………………………………………………… 133
5.7.4 单选列表对话框 ………………………………………………… 134
5.7.5 复选列表对话框 ………………………………………………… 135
5.7.6 自定义对话框 …………………………………………………… 136
5.7.7 进度对话框 ……………………………………………………… 137

5.7.8　自定义进度对话框 ……………………………………………………… 138
5.8　用 Fragment 分割用户界面 …………………………………………………………… 139
　　5.8.1　Fragment 的生命周期 ……………………………………………………… 139
　　5.8.2　设计基于 Fragment 的应用 ………………………………………………… 140
　　5.8.3　Android 支持包 ……………………………………………………………… 148
习题 5 …………………………………………………………………………………………… 149

第 6 章　Android 多媒体 …………………………………………………………………… 151

6.1　Android 的图形绘制 …………………………………………………………………… 151
　　6.1.1　Canvas …………………………………………………………………………… 151
　　6.1.2　Paint ……………………………………………………………………………… 153
　　6.1.3　温度计绘图案例 ……………………………………………………………… 155
　　6.1.4　Bitmap …………………………………………………………………………… 159
　　6.1.5　Matrix …………………………………………………………………………… 161
　　6.1.6　图片缩放功能案例 …………………………………………………………… 161
6.2　Android 多媒体基础 …………………………………………………………………… 164
　　6.2.1　基本类 ………………………………………………………………………… 164
　　6.2.2　权限声明 ……………………………………………………………………… 164
　　6.2.3　Android 多媒体核心 OpenCore ……………………………………………… 165
　　6.2.4　MediaPlayer 类 ………………………………………………………………… 166
6.3　音频播放 ………………………………………………………………………………… 168
　　6.3.1　播放本地资源 ………………………………………………………………… 168
　　6.3.2　播放内部资源 ………………………………………………………………… 168
　　6.3.3　播放网络资源 ………………………………………………………………… 169
6.4　简单音乐播放器案例 …………………………………………………………………… 170
　　6.4.1　案例功能描述 ………………………………………………………………… 170
　　6.4.2　案例程序结构 ………………………………………………………………… 170
　　6.4.3　案例的实现步骤和思路 ……………………………………………………… 170
　　6.4.4　案例参考代码 ………………………………………………………………… 171
　　6.4.5　案例运行效果 ………………………………………………………………… 176
6.5　视频播放 ………………………………………………………………………………… 176
　　6.5.1　使用 MediaPlayer 和 SurfaceView 播放视频 ………………………………… 176
　　6.5.2　使用 MediaPlayer 和 SurfaceView 播放视频案例 …………………………… 177
　　6.5.3　使用 VideoView 播放视频 …………………………………………………… 182
6.6　实现拍照功能 …………………………………………………………………………… 183
　　6.6.1　使用系统自带的拍照应用程序 ……………………………………………… 183
　　6.6.2　自行开发拍照功能 …………………………………………………………… 184
　　6.6.3　Camera 类使用案例 …………………………………………………………… 184

6.7 Android 动画设计 …… 189
　6.7.1 Android 中的逐帧动画 …… 189
　6.7.2 逐帧动画演示案例 …… 189
　6.7.3 Android 中的补间动画 …… 192
　6.7.4 补间动画演示案例 …… 194
　6.7.5 动画监听事件 …… 198
习题 6 …… 199

第 7 章 Android 传感器 200

7.1 传感器的分类 …… 200
　7.1.1 移动传感器 …… 200
　7.1.2 位置传感器 …… 201
　7.1.3 环境传感器 …… 201
7.2 获取传感器事件 …… 203
7.3 传感器坐标系统 …… 206
7.4 详解各种传感器 …… 206
　7.4.1 加速度计 …… 206
　7.4.2 重力传感器 …… 208
　7.4.3 陀螺仪 …… 209
　7.4.4 线性加速度 …… 209
　7.4.5 方向传感器 …… 209
　7.4.6 地磁场传感器 …… 211
　7.4.7 距离传感器 …… 212
习题 7 …… 213

第 8 章 Android 服务简介 214

8.1 Service 的创建及配置 …… 214
8.2 Service 的分类及生命周期 …… 215
　8.2.1 Service 分类 …… 215
　8.2.2 Service 生命周期 …… 215
8.3 启动和停止 Service …… 217
　8.3.1 本地 Service …… 217
　8.3.2 绑定本地 Service …… 218
　8.3.3 Service 案例 …… 218
习题 8 …… 223

第 9 章　Android 广播简介 …… 224

9.1　Android 广播机制 …… 224
9.2　收发广播 …… 224
9.2.1　发送广播 …… 224
9.2.2　接收广播 …… 225
9.2.3　BroadcastReceiver（广播接收者）注册分类 …… 225
9.2.4　静态注册广播案例 …… 226
9.2.5　动态注册广播案例 …… 228
9.3　系统自带的广播 …… 231
9.4　广播分类 …… 232
9.4.1　正常广播 …… 232
9.4.2　有序广播 …… 232
9.4.3　黏滞广播 …… 232
习题 9 …… 233

第 10 章　Android 的数据持久化 …… 234

10.1　SharedPreferences …… 234
10.1.1　获取 SharedPreferences 对象 …… 234
10.1.2　保存 SharedPreferences …… 235
10.1.3　读取 SharedPreferences …… 235
10.1.4　SharedPreferences 案例 …… 235
10.2　文件存储 …… 236
10.2.1　内部存储 …… 237
10.2.2　外部存储 …… 237
10.2.3　文件存储案例 …… 238
10.3　SQLite 数据库存储 …… 242
10.3.1　SQLite 简介 …… 242
10.3.2　SQLiteOpener …… 243
10.3.3　数据库操作 …… 243
10.3.4　SQLite 案例 …… 244
10.4　ContentProvider …… 250
10.4.1　ContentProvider 简介 …… 250
10.4.2　访问手机数据信息 …… 250
10.4.3　ContentProvider 案例 …… 251
习题 10 …… 260

第 11 章 Android 网络编程 ········· 262

- 11.1 URL 统一资源定位符 ········· 262
- 11.2 使用 URLConnection 接口 ········· 262
- 11.3 案例 URLConnection ········· 263
  - 11.3.1 案例功能描述 ········· 263
  - 11.3.2 案例程序结构 ········· 263
  - 11.3.3 案例的实现步骤和思路 ········· 263
  - 11.3.4 案例参考代码 ········· 263
  - 11.3.5 案例运行效果 ········· 265
- 11.4 使用 HttpClient 接口 ········· 266
- 11.5 案例 HttpClient 接口 ········· 266
  - 11.5.1 案例功能描述 ········· 266
  - 11.5.2 案例程序结构 ········· 266
  - 11.5.3 案例的实现步骤和思路 ········· 267
  - 11.5.4 案例参考代码 ········· 267
  - 11.5.5 案例运行效果 ········· 282
- 习题 11 ········· 282

第 12 章 Android 管理器 ········· 283

- 12.1 电话管理器 ········· 283
- 12.2 案例 TelephonyManager ········· 283
  - 12.2.1 案例功能描述 ········· 283
  - 12.2.2 案例程序结构 ········· 283
  - 12.2.3 案例的实现步骤和思路 ········· 283
  - 12.2.4 案例参考代码 ········· 284
  - 12.2.5 案例运行效果 ········· 286
- 12.3 短信管理器 ········· 287
- 12.4 案例 SmsManager ········· 287
  - 12.4.1 案例功能描述 ········· 287
  - 12.4.2 案例程序结构 ········· 287
  - 12.4.3 案例的实现步骤和思路 ········· 287
  - 12.4.4 案例参考代码 ········· 288
  - 12.4.5 案例运行效果 ········· 290
- 习题 12 ········· 290

第13章 LBS 定位服务 ……… 291

13.1 LBS 简介 ……… 291
13.2 LBS 服务模式 ……… 292
　　13.2.1 社交网络和游戏模式 ……… 292
　　13.2.2 生活信息服务模式 ……… 293
　　13.2.3 电子商务模式 ……… 293
13.3 获取位置信息 ……… 294
13.4 百度地图使用案例 ……… 294
　　13.4.1 案例概述 ……… 294
　　13.4.2 案例分析 ……… 295
　　13.4.3 案例实现 ……… 295
习题 13 ……… 305

第14章 综合项目之手机监控 ……… 306

14.1 项目功能需求分析 ……… 306
14.2 应用程序结构设计 ……… 307
14.3 应用程序界面设计 ……… 308
　　14.3.1 欢迎界面布局设计 ……… 308
　　14.3.2 主功能界面布局设计 ……… 309
　　14.3.3 ListView 列表项 Item 布局 ……… 312
14.4 Activity 类设计 ……… 313
　　14.4.1 欢迎界面 Activity ……… 313
　　14.4.2 主功能界面 Activity ……… 316
14.5 应用程序主要功能逻辑设计 ……… 322
　　14.5.1 服务类 SMSService ……… 322
　　14.5.2 获取定位信息类 MyLocationListener ……… 329
14.6 工具类设计 ……… 330
　　14.6.1 缓存类 AppContext ……… 330
　　14.6.2 动作工具类 ActionUtils ……… 331
习题 14 ……… 333

参考文献 ……… 334

第 1 章

Android 概述

本章将对 Android 系统进行总体概述,主要介绍 Android 系统的发展史、系统特性、系统架构及 Android 的开发环境搭建。

1.1 Android 的发展史

1.1.1 Android 的起源

Android(中文名安卓)是一个以 Linux 内核为基础的半开源的移动设备操作系统,可以用在手机、平板电脑和其他移动嵌入式设备上,由 Google 领导的 OHA(Open Handset Alliance,开放手机联盟)维护和开发。Android 的 1.0 Beta 版在 2007 年上市,截至 2012 年底,Android 系统已经成为世界上市场份额最大的手机操作系统。

Android 系统最早由被称为"Android 之父"的 Andy Rubin 带领的一个团队于 2003 年 10 月开始研发,他们当时在美国加州成立了一家高科技公司,叫作 Android Inc. (Android 科技公司),专注于移动设备的智能软件开发。Andy Rubin 和团队中的每个成员都是在科技领域有所建树的技术能手,他们在日夜不停地奋战了两年之后,做出了 Android 系统的整体框架,但却遇到了一个令所有新型公司都为之头疼的难题,就是资金问题。这时,一直与 Rubin 保持着良好私人关系的 Google 公司的两位创始人,向 Android 科技公司伸出了寻求合作的橄榄枝。

2005 年 8 月,Google 低调收购了 Android 科技公司,后者成为 Google 旗下的一部分,Andy Rubin 同时出任 Google 公司工程副总裁,继续负责 Android 项目。

2007 年 11 月,Google 与 84 家硬件制造商、软件开发商和电信运营商联合成立开放手机联盟,来共同研发和改进 Android 系统。紧接着,Google 于 2007 年 11 月发布了 Android 的 1.0 Beta 版,并于次年 9 月发布了 1.0 正式版。此时正值诺基亚的 Symbian 系统在世界手机市场上持续称霸,苹果的 iPhone 也开始大受欢迎,Google 适时成立开放手机联盟并且发布 Android 系统,可以说是为之后的赶超之路打下了坚实的基础。

2010 年末的数据显示,仅正式推出两年的 Android 系统在市场上的占有率已经赶超了称霸 10 余年的诺基亚 Symbian 系统,成为全球第一大智能手机操作系统。2012 年 6 月,Google 在 2012 Google I/O 大会上表示,全球市场上已有超过 4 亿部被 Google 认证的 Android 设备被启动,每天约启动一百万台。

1.1.2 当前的主流移动操作系统

Android 作为一款移动操作系统,自然是与其他的移动操作系统有很大的相同之处。移动操作系统相比于桌面操作系统,一般要注重移动性、个性化、多平台支持和网络连通性。下面来看一下当前比较流行的几款移动操作系统及其各自的特点。

1. Windows Mobile/Phone

它是由微软公司推出的移动设备操作系统。随着 Windows 8 的问世,微软在移动市场上开始发力反击。它的一大优势是将用户熟悉的 Windows 桌面环境应用在了移动设备中,这样可以减少用户的适应时间,并能让用户在移动设备上使用到与桌面 Windows 中相同的应用程序。与其桌面操作系统相同,Windows Mobile/Phone 也是不开放源代码的。Windows Mobile/Phone 使用 C♯ 和 C++ 作为应用开发语言。

2. iOS

它是由苹果公司为 iPhone、iPad 和 iPod Touch 开发的移动操作系统,它的原名叫作 iPhone OS,苹果公司于 2010 年 6 月的 WWDC 大会上宣布将其改为 iOS。iOS 的 1.0 版本于 2007 年 6 月发布,截至目前的最新版是 iOS 9,于 2015 年 6 月在 WWDC 大会上发布。

iOS 操作系统下的游戏和动画程序使用了苹果开发的内置加速器,从而可以获得非常出色的 2D 和 3D 画面效果,同时 iOS 的桌面环境也很美观。与微软的移动操作系统相似,iOS 也是不开源的。iOS 使用 Objective-C 作为应用开发语言。

3. Symbian

大名鼎鼎的 Symbian 操作系统曾经一度称霸手机领域长达数年之久,它的第一代系统 Symbian 5.0 于 1999 年被推出。近些年,Symbian 由于代码滞后、第三方开发难度大、触屏体验不佳和版本兼容性差等缺点,与竞争对手 iOS 和 Android 相比不再具有优势,从而逐渐被对手抢占了市场份额。Symbian 曾开放过一段时间源代码,但后来又封闭了。Symbian 使用 C++ 作为应用开发语言。

4. BlackBerry

中文名为黑莓(不过 RIM 官方一直未认可"黑莓"这个中文名),它是由加拿大的 RIM 公司推出的一种移动电子邮件系统终端,其特点是支持推动式电子邮件、移动电话、文字短信、互联网传真、网页浏览及其他无线信息服务。大部分 BlackBerry 设备都具有全键盘输入功能,BlackBerry 手机特别适合于常处理电话、短信和电子邮件业务的商务人群。BlackBerry 使用 Java 作为应用开发语言。

5. Android

Android系统具备一套完整的智能手机需要具备的功能,且是开放源代码的,虽然后来被证明了其只算是开放了部分源代码,属于半开源的系统,但它仍然是一份不可多得的、功能完整的可用于学习移动开发技术的优秀素材。Android使用Java作为主要的应用开发语言,在需要更改Android的底层功能时,需要使用C或C++。

1.1.3 Android的版本

Android项目的创始人Andy Rubin,过去是一名狂热的机器人爱好者,曾自行设计并制作过小机器人,所以Android曾有两个以机器人命名的内部版本代号,分别是Astro(阿童木,1.0正式版)和Bender(发条机器人,1.1版),这两个版本之后,由于商标问题,Google将Android的版本代号由机器人系列改为现在的甜点系列。

Android的版本代号有一定规律,它按照英文字母A、B、C、D的顺序,以此类推命名,现在最新的版本已经到了字母L,叫作Lolipop(棒棒糖),也就是5.0版。Android的版本发布历史如表1-1所示。

表1-1 Android的版本发布历史

版本号及版本名称	发布时间	重要的更新内容
1.0 Beta	2007-11-12	发布Android SDK预览版,供开发者测试使用,并收集用户反馈
1.0 Astro(阿童木)	2008-09-23	发布第一个正式稳定版Android SDK v1.0,Google开放了Android平台的源代码
1.1 Bender(发条机器人)	2009-02	发布了Android SDK v1.1
1.5 Cupcake(纸杯蛋糕) 基于Linux 2.6.27内核	2009-04-30	支持播放和拍摄影片,并上传到Youtube;支持立体声蓝牙耳机;采用WebKit技术的浏览器;大大提高GPS性能;提供屏幕虚拟键盘;Home界面增加音乐播放器和相册;应用程序自动随着智能手机旋转
1.6 Donut(甜甜圈)内核基于Linux 2.6.29	2009-09-15	支持手势;支持CDMA网络;重新设计了Android Market;支持OpenCore2引擎
2.0/2.1 Éclair(松饼) 基于Linux 2.6.29内核	2009-10-26	支持HTML5;制作新的联系人程序;Google Maps升级为3.1.2版;支持Microsoft Exchange;支持蓝牙2.1;支持内置相机闪光灯;改进虚拟键盘;支持数码变焦
2.2(API 8) Froyo(冻酸奶) 基于Linux 2.6.32内核	2010-05-20	支持将软件安装至扩展内存;集成Chrome的V8 JavaScript引擎到浏览器;支持Adobe Flash 10.1;支持USB分享器;支持WiFi热点功能
2.3(API-9,API-10) Gingerbread(姜饼) 基于Linux 2.6.35内核	2010-12-07	支持WXGA的屏幕尺寸;电话簿集成Intent Call功能;支持NFC(近场通信);优化游戏开发支持;新增下载管理员;从YAFFS转变为EXT4文件系统;加入屏幕截图功能;加入Google Talk;修复了UI

续表

版本号及版本名称	发布时间	重要的更新内容
3.0(API-11) 3.1(API-12) 3.2(API-13) Honeycomb(蜂巢) 基于 Linux 2.6.36 内核	2011-02-02	3.X 都是平板电脑上使用的版本；新版的 Gmail；加入 3D 加速处理；加入专为平板电脑设计的界面；支持多核心处理器；优化了 7 寸平板的显示
4.0(API-14,API-15) Ice Cream SandWich (冰激凌三明治) 基于 Linux 3.0.1 内核	2011-10-19	加入 HOLO 主题，并推荐第三方应用使用该主题；相机自带全景模式；大幅改动用户界面
4.1(API-16) 4.2(API-17) 4.3(API-18) Jelly Bean(果冻豆) 基于 Linux 3.0.31 内核	4.1：2012-06-28 4.2：2012-10-30 4.3：2013-07-24	4.1 版的重要更新内容：增加"牛油"性能，让用户体验更加顺畅；加入 Google Now 活动通知功能；加入脱机语言输入；Google Play 中加入电视片和电影的购买；大幅改变用户界面设计；集成更多的 Google 云；不再自带 Flash Player 4.2 版的重要更新内容：支持多用户账户；加入通知中心里的设置键；更新 Google Now；加入手势输入；支持多媒体无线传输 Miracast；加入照片球(球形全景拍摄)功能 4.3 版的重要更新内容：支持多用户登录；"蓝牙低功耗"功能；支持更多缓冲器对象；新版 OpenGL ES 3.0 着色语言；增加多个纹理的支持；多重渲染目标(Multiple Render Targets)；多重采样抗锯齿(MSAA Render To Texture)；使用统一的纹理压缩格式 ETC；增加 TRIM 指令；新增 App Opt 功能(默认隐藏)
4.4(API-19) KitKat(奇巧巧克力棒) 基于 Linux 3.10 内核	2013-09-03	4.4 版的重要更新内容：支持语音打开 Google Now(在主画面说出"OK Google")；在阅读电子书、玩游戏、看电影时支持全屏模式(Immersive Mode)；优化存储器使用，在多任务处理时有更佳工作的表现；新的电话通信功能；旧有的 SMS 应用程序集成至新版本的 Hangouts 应用程序；Emoji Keyboard 集成至 Google 本地的键盘；支持 Google Cloud Print 服务；支持第三方 Office 应用程序直接打开及存储用户在 Google Drive 内的文件，实时同步更新文件；支持低电耗音乐播放；全新的原生计步器；全新的 NFC 付费集成；全新的非 Java 虚拟机运行环境 ART(Android Runtime)；支持 Message Access Profile(MAP)；支持 Chromecast 及新的 Chrome 功能；支持隐闭字幕
5.0(API-21) Lolipop(棒棒糖) 基于 Linux 3.14 内核	2014-06-25	采用全新 Material Design 界面；支持 64 位处理器；全面由 Dalvik 转用 ART(Android Runtime)编译，性能可提升 4 倍；改良的通知界面及新增优先模式；预载省电及充电预测功能；新增自动内容加密功能；新增多人设备分享功能；强化网络及传输连接性；强化多媒体功能；强化"OK Google"功能；改善 Android TV 的支持；提供低视力的设置，以协助色弱人士；改善 Google Now 功能

1.2　Android系统的特性

Android最初是针对手机研发的操作系统，所以它具有一般手机所具有的电话、短信、邮件、多媒体和上网功能。除此之外，Android还兼顾了用户界面体验和娱乐性，所以在2D和3D的开发方面同样提供了强大的API支持，另外还有标准的多点触控功能。以下是Android系统的一些特性。

1. 显示布局

Android操作系统支持更大的分辨率，VGA、2D显示、3D显示都给予OpenGL ES 2.0标准规格，并且支持传统的智能手机。

2. 数据存储

Android操作系统内置SQLite小型关联式资料库管理系统来负责存储数据。

3. 网络

Android操作系统支持所有的网络制式，包括GSM/EDGE、IDEN、CDMA、EV-DO、UMTS、Bluetooth、WiFi、LTE、NFC和WiMAX。

4. 信息

作为原设计给智能手机使用的操作系统，Android操作系统原生支持短信和邮件，并且支持所有的云信息和服务器信息。

5. 浏览器

Android操作系统中内置的网页浏览器基于WebKit内核，并且采用了Chrome V8引擎。在Android 4.0内置的浏览器测试中，HTML 5和Acid 3故障处理中均获得了满分，并且于2.2版本及之后能原生支持Flash。

6. 编程语言支持

虽然Android操作系统中的应用程序大部分都是由Java编写的，但是Android却是以转换为Dalvik executables的文件在Dalvik虚拟机上运行的。由于Android中并不自带Java虚拟机，因此无法直接运行Java程序。不过Android平台上提供了多个Java虚拟机供用户下载使用，安装了Java虚拟机的Android系统可以运行J2ME的程序。

通常可通过在Android SDK(Android软件开发包)中使用Java作为编程语言来开发应用程序，开发者也可以通过在Android NDK(Android Native开发包)中使用C语言或者C++语言来作为编程语言开发应用程序。同时Google还推出了适合初学者编程使用的Simple语言，该语言类似于微软公司的Visual Basic语言。此外，Google还推出了Google App Inventor开发工具，该开发工具可以快速地构建应用程序，方便新手开发者。

7. 媒体支持

Android 操作系统本身支持以下格式的音频/视频/图片媒体：WebM、H.263 和 H.264(in 3GP or MP4 container)、MPEG-4 SP、AMR 和 AMR-WB (in 3GP container)、AAC 和 HE-AAC (in MP4 or 3GP container)、MP3、MIDI、Ogg Vorbis、FLAC、WAV、JPEG、PNG、GIF、BMP。如果用户需要播放更多格式的媒体，可以安装其他第三方应用程序。

8. 流媒体支持

Android 操作系统支持 RTP/RTSP(3GPP PSS、ISMA)的流媒体以及(HTML5 <video>)的流媒体，同时还支持 Adobe 的 Flash。在安装了 RealPlayer 之后，还支持苹果公司的流媒体。

9. 硬件支持

Android 操作系统支持识别并且使用视频/照片摄像头、多点电容/电阻触摸屏、GPS、加速计、陀螺仪、气压计、磁强计、键盘、鼠标、USB Disk、专用的游戏控制器、体感控制器、游戏手柄、蓝牙设备、无线设备、感应和压力传感器、温度计、加速 2D 位块传输(硬件方向、缩放、像素格式转换)和 3D 图形加速。

10. 多点触控

Android 支持本地的多点触摸，在最初的 HTC Hero 智能手机上就有这个功能。该功能是内核级别(为了避免对苹果公司的触摸屏技术造成侵权)。

11. 蓝牙

Android 支持 A2DP、AVRCP、发送文件(OPP)、访问电话簿(PBAP)、语音拨号和发送智能手机之间的联系。同时支持蓝牙键盘、蓝牙鼠标和蓝牙操纵杆(HID)。

12. 多任务处理

Android 操作系统支持本地的多任务处理。

13. 语音功能

除了支持普通的电话通话之外，Android 操作系统从最初的版本开始，就支持使用语音操作来使用 Google 进行网页搜索等功能。而从 Android 2.2 开始，语音功能还可以用来输入文本、语音导航等。

14. 无线共享功能

Android 操作系统支持用户使用本机充当"无线路由器"，并且将本机的网络共享给其他智能手机，其他机器只需要通过 WiFi 查找到共享的无线热点，就可以上网。而在

Android 2.2 版本之前的操作系统,则需要通过第三方应用程序或者其他定制版系统来实现这个功能。

15. 截图功能

从 Android 4.0 版本开始,Android 操作系统便支持截图功能。该功能允许用户直接抓取智能手机屏幕上的任何画面,用户还可以通过编辑功能对截图进行处理,还可以通过蓝牙/E-mail/微博/共享等方式发送给其他用户或者上传到网络上,也可以复制到计算机中。

16. 跨平台

由于 Android 操作系统的开放性和可移植性,它可以被用在大部分电子产品上,主要包括智能手机、上网本、平板电脑、个人电脑、笔记本电脑、电视、机顶盒、MP3 播放器、MP4 播放器、掌上游戏机、家用主机、电子手表、电子收音机、耳机、汽车设备、导航仪、CD 机、VCD/DVD 机等设备。

Android 操作系统大多搭载在使用了 ARM 架构的硬件设备上。但是同样也有支持 X86 架构的 Android 操作系统,例如 Google 的 Google TV 就是使用一个特别定制的 X86 架构版本的 Android 操作系统。

同样,苹果公司的 iOS 设备,例如 iPhone、iPod Touch 以及 iPad 产品都可以安装 Android 操作系统,并且可以通过双系统启动工具 OpeniBoot 或者 iDroid 来运行 Android 操作系统。微软的 Windows Mobile、Windows Phone 产品也一样可以。另外 Android 亦已成功移植到搭载 WebOS 系统的 HP TouchPad 以及搭载 Meego 系统的 Nokia N9 等设备。

17. 应用程序的安全机制

Android 操作系统使用了沙箱(Sandbox)机制,所有的应用程序都会先被简单地解压缩到沙箱中进行检查,并且将应用程序所需的权限提交给系统,将其所需权限以列表的形式展现出来,供用户查看。例如一个第三方浏览器需要"连接网络"的权限,或者一些软件需要拨打电话、发送短信等权限。用户可以根据权限来考虑自己是否需要安装,用户只有在同意了应用程序权限之后,才能进行安装。

1.3 Android 系统架构

Android 是在 Linux 内核的基础上,使用一种可称为"软件层级"的架构组织起来的。"软件层级"架构是指它含有多个层次,而每层都是由多个软件模块或软件库组成的。Android 的架构共有 4 层,如图 1-1 所示。

纵观整个 Android 系统架构,各种开源的软件包和各种主流的编程语言全部都有"用武之地",从下到上,一同构建出了一款移动操作系统。从编程语言的角度来看,图 1-1 中所有红色部分都是 C 语言写成的;所有绿色部分都是由 C++ 为主,而辅之以 C 写成有用的本地库;所有蓝色部分则都是Java语言写成的。从开发者的角度来看,如果只是开发一般的应用程序,则只需要使用Java语言在应用层做开发即可。如果要开发一些个人或公司自用的框架,则同样使用Java在前两层进行开发即可。如果要做Android系统级开发,则需要深入本地库和Java运行时环境层,使用C++和C进行开发。如果需要开发Android的驱动程序,则需要从Linux内核层开始开发。

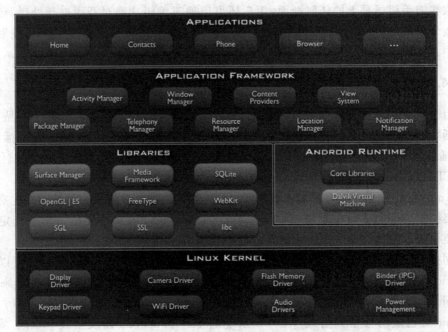

图1-1　Android系统架构示意图

1.3.1　Android应用层

应用层是Android设备真正与用户进行交互的一层,Android设备上的应用程序都是运行在这一层的。其中包括Google开发的应用软件,例如电话、短信、电子邮件、浏览器等,也包括一般开发者所开发的应用软件。这一层使用Java作为其开发语言。

1.3.2　Android应用框架层

框架层是从事Android开发的基础,很多核心应用程序也是通过这一层来实现其核心功能的。该层简化了组件的重用,开发人员可以直接使用其提供的组件来进行快速的应用程序开发,也可以通过继承而实现个性化的拓展。由于在其下已经运行了Java运行时环境,所以这一层使用Java语言作为开发语言。Android应用框架层主要包含如下内容。

1. Activity Manager(活动管理器)

管理各个应用程序生命周期以及通常的导航回退功能。

2. Window Manager（窗口管理器）

管理所有的窗口程序。

3. Content Provider（内容提供器）

使得不同应用程序之间存取或者分享数据。

4. View System（视图系统）

构建应用程序的基本组件。

5. Notification Manager（通告管理器）

使得应用程序可以在状态栏中显示自定义的提示信息。

6. Package Manager（包管理器）

Android 系统内的程序管理。

7. Telephony Manager（电话管理器）

管理所有的移动设备功能。

8. Resource Manager（资源管理器）

提供应用程序使用的各种非代码资源，如本地化字符串、图片、布局文件、颜色文件等。

9. Location Manager（位置管理器）

提供位置服务。

10. GTalk Service

提供 Google Talk 服务。

1.3.3　Android 核心库

本地库是应用程序框架的基础，是连接应用程序框架层与 Linux 内核层的重要纽带。主要含有以下几个重要的库。

1. Surface Manager

执行多个应用程序时，负责管理显示与存取操作间的互动，另外也负责 2D 绘图与 3D 绘图的显示合成。

2. Media Framework

多媒体库,基于 PacketVideo OpenCore,支持多种常用的音频、视频格式录制和回放,编码格式包括 MPEG4、MP3、H.264、AAC、ARM。

3. SQLite

小型的关系型数据库引擎。

4. OpenGL|ES

根据 OpenGL ES 1.0 API 标准实现的 3D 绘图函数库。

5. FreeType

提供点阵字与向量字的描绘与显示。

6. WebKit

一套网页浏览器的软件引擎。

7. SGL

底层的 2D 图形渲染引擎。

8. SSL

在 Android 上通信过程中实现握手。

9. Libc

从 BSD 继承来的标准 C 系统函数库,专门为基于嵌入式 Linux 的设备定制。

1.3.4 Android 运行时

Android 应用程序是用 Java 语言编写的,所以 Android 需要一个 Java 的运行时环境,该环境又包括核心库和 Dalvik 虚拟机两部分。

核心库提供了 Java 语言 API 中的大多数功能,同时也包含了 Android 的一些核心 API,如 android.os、android.net、android.media 等。

Android 程序不同于 J2ME 程序。每个 Android 应用程序都有一个专有的进程,并且不是多个程序运行在一个虚拟机中,而是每个 Android 程序都有一个 Dalvik 虚拟机的实例,并在该实例中执行。Dalvik 虚拟机不是传统的基于栈的虚拟机,而是一种基于寄存器的 Java 虚拟机,并进行了内存资源使用的优化以及支持多个虚拟机的特点。需要注意的是,不同于 J2ME,Android 程序在虚拟机中执行的并非编译后的字节码,而是通过转换工具 dx 将 Java 字节码转成 dex 格式的中间码。

1.3.5 Linux 内核层

Android 是在 Linux 内核的基础上构建的,Android 的内核属于 Linux 内核的一个分支,它并不是 GNU/Linux,因为一般在 GNU/Linux 中被支持的功能,在 Android 大多没有被支持。众所周知,Linux 是一个开源的操作系统,由非营利的组织——Linux 基金会所管理。虽然 Linux 是开源的,但是 Android 必须在 GNU GPL(用于保护开源软件的一个授权规范)的许可下使用 Linux 的源码,才可以商用。所以为了达到商业应用的目的,Android 必须去除被 GNU GPL 所约束的部分。Android 去除了 Cairo、X11、Alsa、FFmpeg、GTK、Pango 和 Glibc 等,并以 Bionic 取代 Glibc、以 Skia 取代 Cairo、以 Opencore 取代 FFmpeg 等。Android 并没有用户空间驱动,而是将所有的驱动都放在内核空间中,并以 HAL 隔开版权问题。

目前,Android 的 Linux 内核层包括安全管理、内存管理、进程管理、网络协议栈、驱动程序模型和电源管理等,这些都依赖于 Linux 内核。由于 Linux 内核全部使用 C 语言编写,所以 Android 的 Linux 内核层也全部是用 C 语言编写的。

1.4 Android 环境搭建

Android 可以使用 Eclipse 作为开发环境,代码编写、程序调试和运行等都可以在 Eclipse 中完成。使用 Eclipse 开发 Android 应用程序,首先要在 Eclipse 中安装 ADT (Android Developer Tools)插件,然后对 Android SDK 进行配置。Android 开发环境中的所有工具都可以免费下载和使用。

本书中介绍的是在 Windows 平台上搭建 Android 开发环境的方法,有关在 Linux 和 Mac OS 平台上的搭建方法,均可以参考本文。

1.4.1 Java 开发环境安装与配置

1. 安装 JDK

从 Java 的官网 http://www.oracle.com/technetwork/java/javase/downloads/index.html 上下载 JDK,然后直接安装,注意要将 JDK 和 JRE 都要安装,且将两者置于同一个目录下,如图 1-2 所示。

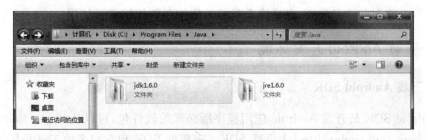

图 1-2　安装好的 JDK 文件夹示意图

安装好 JDK 之后,要按照如下步骤对 JDK 进行配置。

(1) 在 Windows 的系统环境变量中,新建环境变量名 JAVA_HOME,其值为 JDK 安装路径,如图 1-3 所示。

(2) 在 PATH 环境变量中添加％JAVA_HOME％\bin。

图 1-3　JAVA_HOME 环境变量设置

(3) 新建环境变量名 CLASSPATH,变量值设置为:

.;%JAVA_HOME%\lib\dt.jar;%JAVA_HOME%\lib\tools.jar

(4) 可以在命令行(CMD)中输入 java-version,如果有正常的输出结果,就表明 JDK 已安装并配置完成,如图 1-4 所示。

```
C:\Users\Administrator>java -version
java version "1.7.0_45"
Java(TM) SE Runtime Environment (build 1.7.0_45-b18)
Java HotSpot(TM) 64-Bit Server VM (build 24.45-b08, mixed mode)

C:\Users\Administrator>
```

图 1-4　JDK 测试

2. 安装 Eclipse

Eclipse 最初由 OTI 和 IBM 两家公司的 IDE 产品开发组于 1999 年 4 月创建。它的设计思想是"一切皆插件",它本身是一个免费并开源的项目,但其上的插件可能是收费的。它本身只能作为 Java 的开发环境,可以开发 J2SE、J2EE、EJB 和 JSP Web 项目,而且具有非常强大的扩展性,可以通过安装相应的插件来作为 C、C++、Python、PHP、Perl、Ruby 等多种语言的开发环境。另外,Eclipse 还是跨平台的,也常用在 Linux 上,Linux 上的许多 C 和 C++ 的开发任务都可以使用 Eclipse 完成,例如想用 Eclipse 开发 C 和 C++,安装 CDT 即可。近几年来,包括 Oracle 在内的许多大公司都加入到 Eclipse 项目中,并宣称 Eclipse 将会支持任何语言的开发,成为 IDE 的集大成者。有关 Eclipse 的详细信息可以在其官网 http://www.eclipse.org 上查询。

Eclipse 可以从其官网 http://www.eclipse.org/downloads/ 上下载。Eclipse 每年 6 月份会更新一次新版本,并会在 9 月和次年 2 月发布该版本对应的 SR1 和 SR2 版本,下载 Eclipse 最新版即可。

Eclipse 是不需要安装的,直接将下载好的软件包解压缩即可使用。

1.4.2　安装 Android 开发环境

1. 下载 Android SDK

Android SDK 是开发 Android 应用程序所必需的软件包,可以从 http://developer.android.com/sdk/index.html 上下载 SDK。完整的 SDK 包含很多的工具和从 Android

1.5 到最新版的所有包，共有几个 GB 的大小。从网上直接下载的 SDK 只包含最核心的 SDK Tools，它比较小，便于下载，而通过它可以使用自动更新的方式下载到其他所有版本的 SDK 资源。

下载好基本的 SDK 包后，直接解压缩，得到一个 Android SDK 文件夹，这个 SDK 文件夹要用于之后的 ADT 配置。由于 Android SDK 是可以不断更新的，所以在整个 Android 的开发中都会被使用到。注意不要更改这个文件夹的名称，只需要保持压缩包的名称即可，且它的文件路径中最好不要出现中文和空格。

Android 采用自动更新的方式帮助开发者管理 SDK。SDK 文件夹内会包含一个可执行文件，叫作 SDK Manager.exe，单击它会出现如图 1-5 所示的 Android SDK Manager 界面，它会自动连接 Google 的 Android 服务器，检查最新的 SDK 版本（图 1-5 中显示该版本已被安装）和工具，供用户选择下载并安装。以后就可以使用这个工具更新 SDK 版本和开发工具。在 Eclipse 配置好 ADT 之后，也可以从 Eclipse 中的菜单 Windows->Android SDK Manager 启动 SDK Manager，出现的界面与图 1-5 所示的相同。

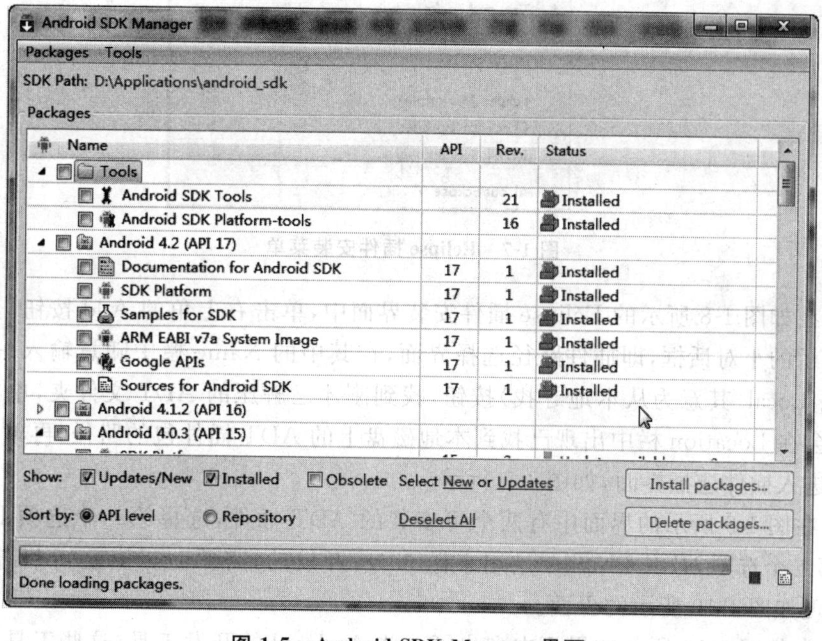

图 1-5 Android SDK Manager 界面

2. 在 Eclipse 中安装并配置 ADT

ADT 是 Android 的开发工具，与 CDT 和 PDT 类似，都是以 Eclipse 的插件形式，安装在 Eclipse 之上，以使 Eclipse 可以开发相应环境下的项目。

1）安装 ADT

ADT 安装有以下两种方式，推荐读者先使用方式二，如果不成功再使用方式一。方式二简单，但偶尔会找不到网络资源，而方式一是不会出现问题的。

方式一,先下载 ADT,然后在本地安装。

(1) 从 Android 的官网 http://developer.android.com/sdk/installing/installing-adt.html 上找到 ADT 的压缩包,下载到本地,如图 1-6 所示。下载 ADT 之后,将其解压缩,并放在无中文、无空格的路径下。

Package	Size	MD5 Checksum
ADT-21.0.0.zip	13556487 bytes	7db4eaae5df6a34fd853317a2bd8250b

图 1-6　ADT 压缩包的下载链接

(2) 在 Eclipse 的顶部菜单栏中选择 Help->Install New Software,如图 1-7 所示。

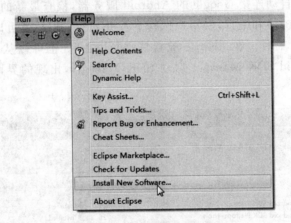

图 1-7　Eclipse 插件安装菜单

(3) 在如图 1-8 所示的 Eclipse 插件安装界面中,单击右上角的 Add 按钮,会弹出如图 1-8 所示的小对话框,即插件路径选择界面,在其中的 Name 栏中随意输入一个名字,然后单击 Local(其意为从本地寻找)按钮,找到刚才已解压的 ADT 文件夹,单击 OK 按钮,这时会在 Location 栏中出现已找到本地磁盘上的 ADT 插件包的路径,再单击 OK 按钮,将会进入插件选择界面,如图 1-9 所示。

(4) 在图 1-9 所示的界面中有两个可安装的 ADT 插件,选择第一个选项 Developer Tools(第二个选项 NDK Plugins 是用来做 NDK 开发的,需要时再安装),单击 Next 按钮,会出现如图 1-10 所示的界面。

(5) 在如图 1-10 所示的界面中列出了一些 Android 的开发工具,这些工具都是很有用的,我们会在下文予以介绍。这里先把它们都选上,单击 Select All 按钮即可完成全选,然后单击 Finish 按钮,之后会出现一个协议接受的界面,选择 Accept,然后单击 Install 按钮,即可开始安装。安装好 ADT 后,重启 Eclipse。至此,ADT 的安装过程结束。

方式二,直接在线安装。Eclipse 提供了直接在线安装插件的功能,可以直接在线安装 ADT,步骤如下。

在 Eclipse 的顶部菜单栏中选择 Help->Install New Software。在如图 1-11 所示的插件安装界面中单击右上角的 Add 按钮,然后会出现插件路径选择对话框。在 Name

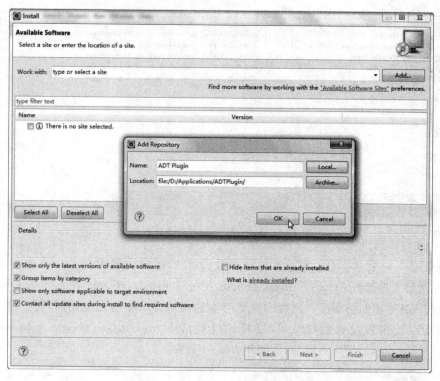

图 1-8　本地安装 ADT 的路径选择

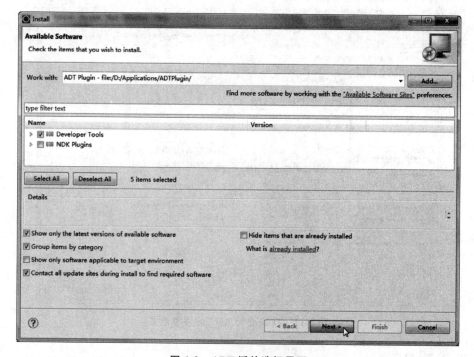

图 1-9　ADT 插件选择界面

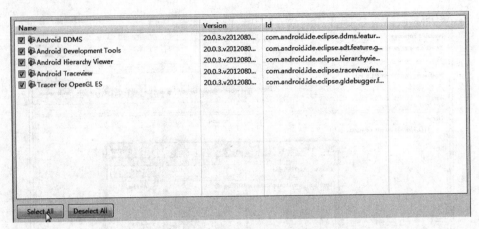

图 1-10　Android 开发工具选择界面

栏中随意输入一个名字(注意不要与已存在的名字相同),然后在 Location 栏中输入 ADT 在网络上的地址 https://dl-ssl.google.com/android/eclipse/,之后单击 OK 按钮,会出现与图 1-9 相似的界面。之后的步骤与本地安装方式中的完全相同,可参考之。如果单击 OK 按钮后没有预期结果,则将地址换成 http://dl-ssl.google.com/android/eclipse/再试一遍,如果再不成功,则使用方式一安装。

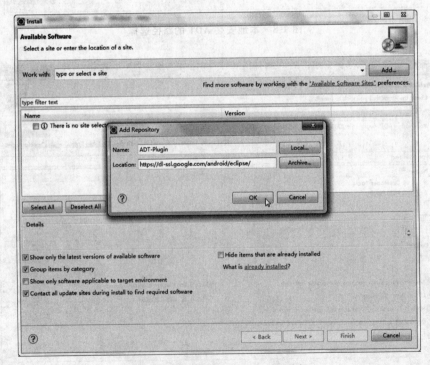

图 1-11　在线安装 ADT 的路径选择

2) 配置 ADT

启动 Eclipse,在菜单栏中选择 Windows—>Preferences,会出现 Eclipse 的配置界

面,Eclipse 的大部分设置都在这里完成。由于已安装好 ADT,所以会在 Eclipse 的配置界面的左边出现一个名为 Android 的条目,单击这个条目,进入到 Android 设置中,如图 1-12 所示。在 SDK Location 栏中设置 Android SDK 的文件夹路径,可以单击 Browse 按钮选择 Android SDK 文件夹,然后单击 OK 按钮,之后单击界面下方的 Apply 按钮,即可完成 ADT 配置。如果 Android SDK 的路径没问题,则会出现如图 1-12 所示的界面,它列出了所有已安装的 SDK 的版本,这表示 ADT 已配置成功。至此,Android 的开发环境就搭建好了,可以开始 Android 的应用开发之旅了。

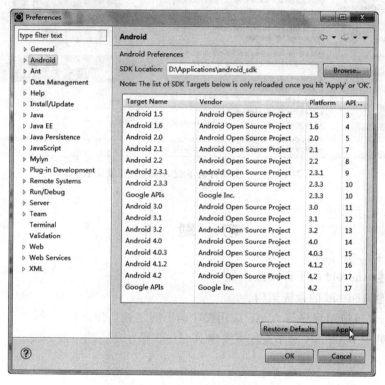

图 1-12 ADT 配置界面

为了方便开发者,Google 在其 Android 官网上提供了打包好的 Android 开发工具包,供开发者下载,地址为 http://developer.android.com/sdk/index.html,如图 1-13 所示。

下载好的包中包含了已安装了 ADT 的 Eclipse、SDK Tools、SDK Platform-tools、Android Support Libraries,以及最新版的 SDK Platform 和该版本对应的建立模拟器所需的系统镜像(Android System Image),这些是可以开发 Android 所必需的最少的工具组合。

下载好软件包之后,解压缩,然后启动 Eclipse,由于 ADT 已被安装在了 Eclipse 之上,且 SDK 也已存在于本地磁盘中,所以只需要配置 ADT 即可,具体方法参见本节中配置 ADT 的内容,这里不再赘述。

图 1-13　一次性打包下载 Android 开发工具

习　题　1

1. 简述 Android 的平台特性。
2. 描述 Android 的系统架构。
3. Android 应用程序与 J2ME 应用程序的不同之处是什么？
4. 说明在 Windows 操作系统下搭建 Android 开发环境的一般步骤。

第 2 章 开发工具介绍及项目结构

本章首先介绍 Android 常用的开发调试工具,然后展示如何创建、运行一个简单的 Android 项目,并对 Android 项目的结构进行分析,最后介绍 Android 的 4 大应用程序组件。

2.1 Android 开发工具介绍

2.1.1 adb

adb(Android Debug Bridge,调试桥)是一个 debug 工具,可将其他工具接入模拟器和设备,通过 adb 可以在 Eclipse 中方便地通过 DDMS 来调试 Android 程序。除了可以让其他工具(ADT 插件)功能生效以外,还可以使用命令行上传或下载文件、安装或卸载程序包、通过进入设备或模拟器的 shell 环境访问许多其他功能。运行 Eclipse 时 adb 进程就会自动运行。

adb 是一个客户端/服务器端程序,其中客户端是用来操作的计算机,服务器端是 Android 设备。adb 安装在 Android SDK 的 platform-tools 目录下,它的常用命令如下。

1. 查看设备

命令: adb devices

这个命令是查看当前连接的设备,连接到计算机的 Android 设备或者模拟器将会在此列出显示。如图 2-1 所示,在运行 adb devices 命令之后可以看到当前连接的一个设备 emulator-5554 device。

```
D:\Download\adt-bundle-windows-x86_64-20140702\sdk\platform-tools>adb devices
List of devices attached
emulator-5554    device
```

图 2-1 利用 adb 命令查看当前连接的 Android 设备

2. 安装软件

命令: adb install <apk 文件路径>

这个命令将指定的 apk 文件安装到设备上。

3. 卸载软件

命令 1：adb uninstall ＜软件名＞

命令 2：adb uninstall -k＜软件名＞

如果加-k 参数，则为卸载软件但是保留配置和缓存文件。

4. 登录设备 shell

命令 1：adb shell

命令 2：adb shell ＜command 命令＞

这个命令将登录设备的 shell。

后面加＜command 命令＞将直接运行设备命令，相当于执行远程命令。

5. 从计算机上发送文件到目标机

命令：adb push ＜本地路径＞ ＜远程路径＞

用 push 命令可以把计算机上的文件或者文件夹复制到 Android 目标机中。

6. 从目标机上下载文件到计算机

命令：adb pull ＜远程路径＞ ＜本地路径＞

用 pull 命令可以把 Android 目标机上的文件或者文件夹复制到计算机。

7. 显示帮助信息

命令：adb help

2.1.2 AVD

AVD(Android Virtual Device,安卓虚拟设备)，一般称其为 Android 模拟器，可以用来模拟一个 Android 手机或平板电脑，由于可以虚拟出各个 API 版本、各种屏幕分辨率的 Android 设备，所以 AVD 在 Android 开发中很常用。

AVD 没有 SIM 卡，也没有 WiFi 网络，硬件资源受限，所以有部分 Android 应用程序需要在真机(真实的 Android 设备)上调试，例如使用到 3G 和 WiFi 网络、使用 3D 渲染、使用到 SIM 卡等功能的应用程序。

1. AVD 的创建

建立 AVD 的步骤如下。

第一步，在 Eclipse 中，选择菜单项 Windows—＞Android Virtual Device Manager。

第二步，在出现的 AVD 管理界面中，单击右上角的 New 按钮，就会出现新建 AVD 的界面，在此界面中可以配置 AVD，如图 2-2 所示。配置完成后，单击 OK 按钮，即可创建一个新的 AVD。这个被创建的 AVD,就相当于一个连接在计算机上的 Android 设备，可以用来运行和调试 Android 程序。

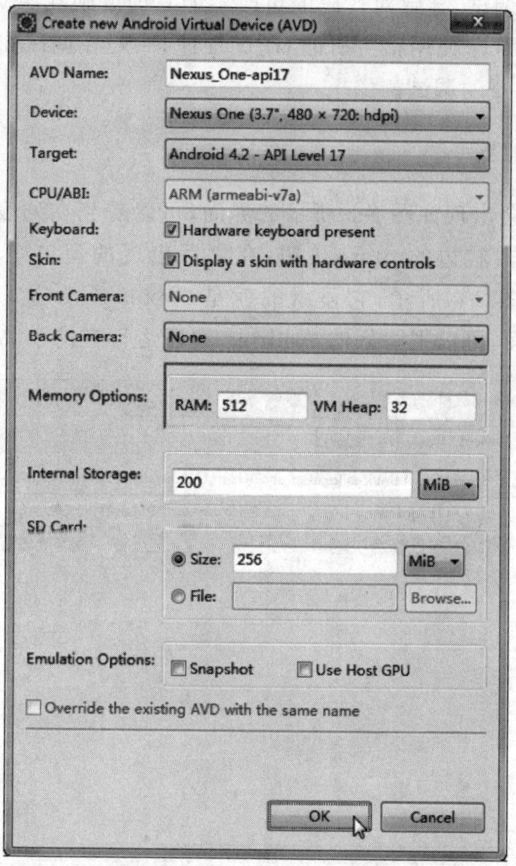

图 2-2 AVD 建立界面

图 2-2 中 AVD 的配置信息说明如下。

（1）AVD Name：AVD 名称，作为标识，开发者自用，能识别不同的 AVD 即可。

（2）Device：要模拟的设备，一般为现在 Google 的 Nexus 系列手机，以及其他手机和平板电脑，区别在于屏幕分辨率不同，可以建立多个 AVD 来做屏幕适配。

（3）Target：模拟器的 API 版本。

（4）Keyboard/Skin：键盘布局/皮肤设置，按默认选中即可。

（5）RAM：模拟器的内存大小，会共享宿主机的内存，一般为默认大小即可，太大会降低宿主机的速度，太小会降低模拟器的速度。

（6）VM Heap：模拟器堆栈大小，默认值即可。

（7）Internal Storage：模拟器 ROM 大小，存放安装到模拟器上的程序，占用计算机磁盘空间。

（8）SD Card Size：模拟器的 SD Card 大小，一般存放多媒体、照片等大文件，占用计算机的磁盘空间。与 ROM 占用计算机空间的机制不同，只要 AVD 建立好，就会占用这样大小的磁盘空间，所以这个值不宜设置得太大，不够用时再设置大一些即可。

(9) Snapshot：快照功能，会保存上次的 AVD 显示界面，以加快下一次 AVD 的启动，但有时会影响程序调试，导致对代码的更改不能立即反映在 AVD 的界面上。

(10) Use Host GPU：使用宿主机的 GPU 加速，一般在调试 3D 游戏时开启，不过通常都使用真机做 3D 游戏的调试。

2. 启动 AVD 模拟器

建好 AVD 后，就会出现如图 2-3 所示的界面，可以看到在列表中出现了刚才建立的 AVD，选择它，然后单击右边的 Start 按钮，在之后出现的窗口中单击 Launch 按钮，该 AVD 就被启动了，如图 2-4 所示，它显示的就是 Nexus_One 这款手机在 Android 4.2（API 17）版本下的界面，可以把写好的 Android 程序运行在这个 AVD 上。

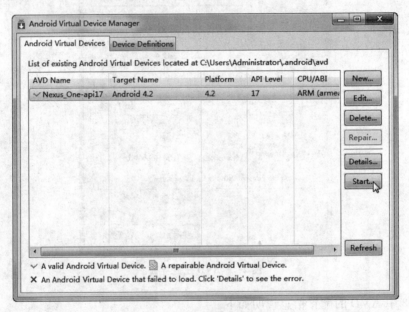

图 2-3　AVD 选择界面

2.1.3　DDMS

DDMS (Dalvik Debug Monitor Service)是 Android 开发环境中的 Dalvik 虚拟机调试监控服务。DDMS 提供了测试设备截屏，针对特定的进程查看正在运行的线程及堆信息、Logcat、广播状态信息、模拟电话呼叫、接收 SMS、虚拟地理坐标等功能。

1. DDMS 的启动方法

(1) DDMS 安装在 Android SDK 的 tools 目录下，可以在此目录下双击 ddms.bat 直接运行；

(2) 在 Eclipse 中，选择菜单 Window->Open Perspective，双击菜单项 DDMS，也可启动 DDMS，DDMS 界面如图 2-5 所示。

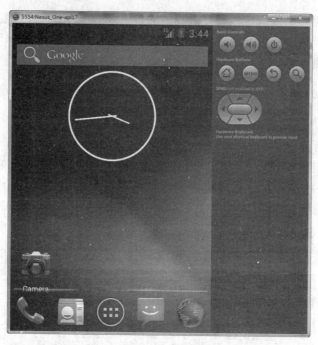

图 2-4 AVD 启动后的初始界面

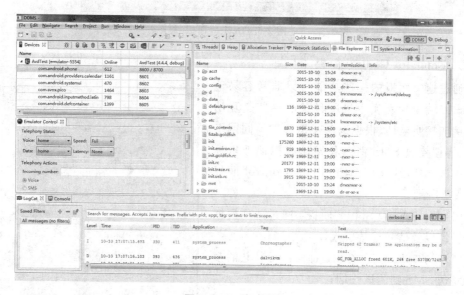

图 2-5 DDMS 界面

2. DDMS 工作原理

DDMS 将搭建起 IDE 与测试终端(Emulator 或 Connected Device)的链接。DDMS 可以实时监测到测试终端的连接情况,当有新的测试终端连接后,DDMS 将捕捉到终端的 ID,并通过 adb 建立调试器,从而实现发送指令到测试终端的目的。

DDMS 监听第一个终端应用程序进程的端口为 8600,应用程序进程将分配 8601,如果有更多终端或者更多应用程序进程将按照这个顺序依次类推。DDMS 通过 8700 端口接收所有终端的指令。

3. DDMS 的功能

可以通过 DDMS 的图形界面了解到更多 DDMS 的功能。

1) Devices 面板

在这个面板可以看到所有与 DDMS 连接的终端的信息,以及每个终端正在运行的应用程序进程,每个进程的右边相对应的是与调试器链接的端口。因为 Android 是基于 Linux 内核开发的平台,同时也保留了 Linux 中特有的进程 ID,它介于进程名和端口号之间。Devices 面板如图 2-6 所示。

图 2-6 Devices 面板

2) Emulator Control 面板

通过这个面板的一些功能可以非常容易地使测试终端模拟真实手机所具备的一些交互功能。例如接听电话、根据选项模拟各种不同网络情况、模拟接收 SMS 消息和发送虚拟地址坐标用于测试 GPS 功能等。Emulator Control 面板如图 2-7 所示。

(1) Telephony Status 面板:通过选项模拟语音质量以及信号连接模式;

(2) Telephony Actions 面板:模拟电话接听和发送 SMS 到测试终端;

(3) Location Control 面板:模拟地理坐标或者模拟动态的路线坐标变化并显示预设的地理标识。可以通过 Manual 面板手动为终端发送二维经纬坐标;或在 GPX 面板通过 GPX 文件导入序列动态变化地理坐标,从而模拟行进中 GPS 变化的数值;在 KML 面板可以通过 KML 文件导入独特的地理标识,并以动态形式根据变

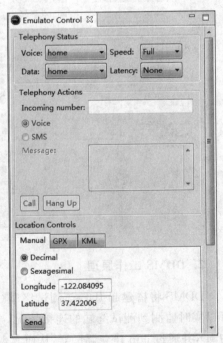

图 2-7 Emulator Control 面板

化的地理坐标显示在测试终端。

3）其他功能

Devices 面板右边的窗口中还有 Threads、Heap、Allocation Tracker、File Explorer、Network Statistics 和 System Information 选项卡，如图 2-8 所示，可以显示线程统计信息、栈信息、内存分配跟踪情况、Android 设备的文件系统、网络使用统计信息和 Android 设备的系统信息。

图 2-8 其他功能选项卡

在这些选项卡中，最常使用的是 File Explorer 选项卡，它可以把文件上传到 Android 设备，或从 Android 设备下载文件，也可以进行文件删除操作。

4）LogCat 面板

在 DDMS 界面下方的是 LogCat 面板，可以输出 AVD 或 Android 设备的一些运行和调试信息。LogCat 是 Android 日志系统的名称，是一个日志记录工具，可以通过 Eclipse、adb 读取 LogCat 数据，它可以提供系统中相关事件的诊断信息。开发者可以由此将应用程序的调试和诊断信息发送到 LogCat。这个工具很常用，可以理解为 C 或 Java 程序中的控制台输出，常被用来在手写代码的调试方式中，输出调试信息，使用时在代码中先使用 import 关键字导入 android.util.Log 包，然后在代码中使用 Log.d (String,String)、Log.i(String,String)等方法输出调试信息。

LogCat 面板如图 2-9 所示，其右边窗格显示的是 LogCat 日志，左边窗格可以过滤一些调试信息。

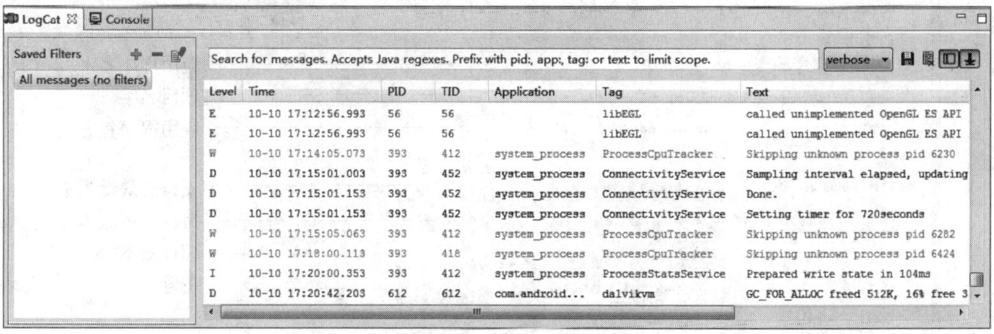

图 2-9 LogCat 面板

2.2 创建第一个 Android 项目

Android 的开发环境搭建好之后，就可以开发并运行 Android 程序了。Android 程序是运行在手机等移动设备上的，但是开发 Android 程序却一般不会放在移动设备上完成,而是在安装有 Windows、Linux 或 Mac OS 的个人计算机(PC)上完成。这一过程与传统的嵌入式开发一样,有宿主机和目标机之分。在本书中,把开发 Android 程序的机器叫作宿主机或 PC,把运行 Android 程序的机器叫作目标机、目标设备或 Android 移动设备,它可以是 Android 手机、Android 平板或 AVD。

如果有 Android 移动设备,可以在 PC 上编写 Android 代码,然后将移动设备与 PC 相连,之后就可以使用移动设备来运行、调试 Android 程序;如果没有 Android 移动设备,可以使用 AVD 进行 Android 程序的运行与调试。

2.2.1 如何创建 Android 项目

创建一个最简单的 Android 应用程序的步骤如下。

1. 建立项目

在 Eclipse 中选择菜单 File->New->Projects,在其中双击菜单项 Android Application Project,之后会出现"新建 Android 项目"对话框,如图 2-10 所示,在其中输入应用程序名字、项目名称、应用程序包名,选择 SDK 支持的最低版本、目标版本、编译版本和主题之后,单击 Next 按钮,之后会出现"项目配置"对话框,如图 2-11 所示,在此界面中默认所有选项,单击 Next 按钮,出现如图 2-12 所示的"项目图标配置"对话框,选择默认,再次单击 Next 按钮,出现如图 2-13 所示的"新建 Activity 选择"对话框,选择默认,再单击 Next 按钮,出现如图 2-14 所示的"新建 Activity 配置"对话框,输入入口 Activity 及其界面布局文件名称后,最后单击 Finish 按钮,Android 项目就创建好了。

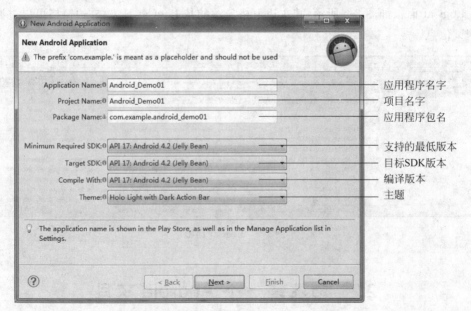

图 2-10 "新建 Android 项目"对话框

第 2 章 开发工具介绍及项目结构　27

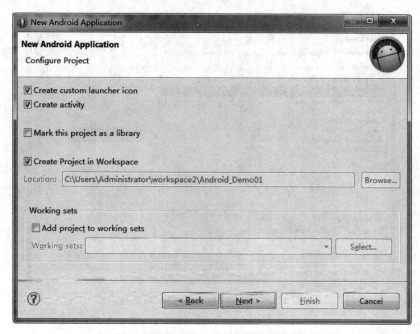

图 2-11 "项目配置"对话框

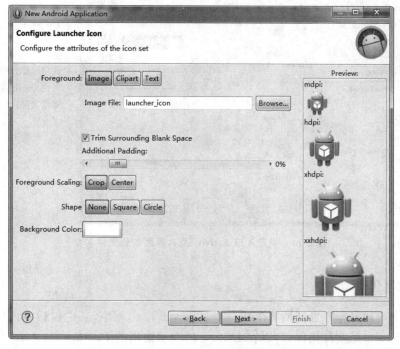

图 2-12 "项目图标配置"对话框

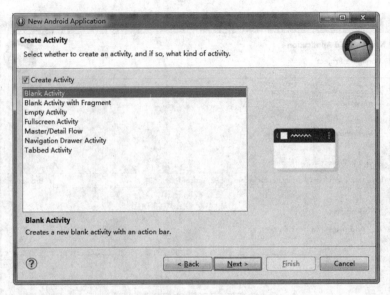

图 2-13 "新建 Activity 选择"对话框

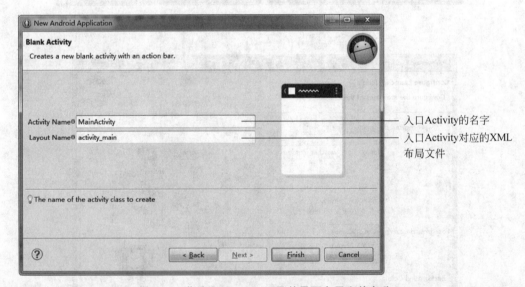

图 2-14 指定入口 Activity 及其界面布局文件名称

2. 编写代码

通过上面的操作开发环境将生成一个 Android 应用程序,其中包括了入口 Activity 和对应的界面布局文件代码,可以在 Eclipse 的左侧窗格的 Package Explorer 面板中,通过双击某个项目条目,来打开源文件进行代码编写或修改。程序编辑窗口如图 2-15 所示。

在上面新建的第一个项目中,可以修改任何代码,直接运行这个项目。

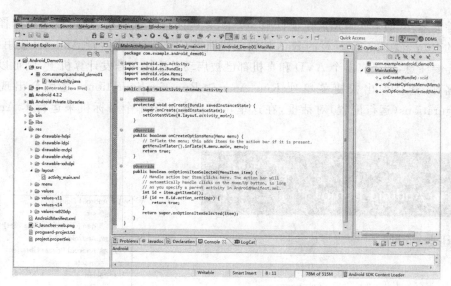

图 2-15 Android 程序编辑窗口

3. 运行 Android 应用程序

在 Package Explorer 面板中的 Android_Demo01 项目上右击，在弹出的快捷菜单中选择 Run As—>Android Application，即可自动地在已开启的 AVD 上运行这个项目，如图 2-16 所示。运行后的效果如图 2-17 所示，从图中可以看出该 Android 程序实现了在界面上显示一行文本信息"Hello world!"。

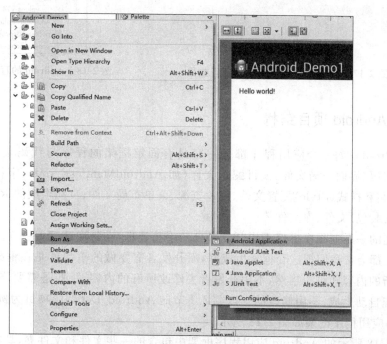

图 2-16 运行 Android 项目

除此之外,如果此时焦点在项目中 src/目录下的 Java 文件中,还可以通过选择菜单栏中的 Run->Run、工具栏的 Run 按钮或快捷键 Ctrl+F11 来运行 Android 应用程序。

需要注意,如果有多个 AVD 和真机都已被启动且连接在开发计算机上,则可以右击项目名称,选择弹出快捷菜单项 Run As->Run Configurations,之后会出现 Run Configurations(运行配置)对话框,在这里可以指定使用哪个目标机来运行本 Android 项目。

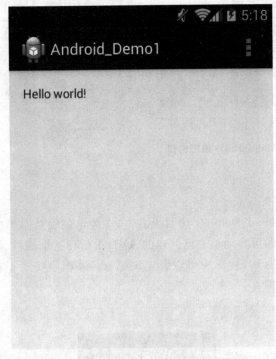

图 2-17 Android_Demo01 项目运行效果

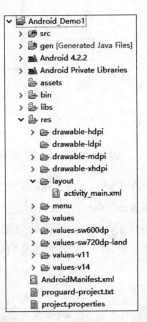

图 2-18 Android 的程序框架示意图

2.2.2 Android 项目结构

Android 的每一个应用程序都是一套具有固定框架的程序与代码集合,里面包括 Java 源代码、界面布局文件、项目配置文件(如 AndroidManifest.xml)、字符串数据配置文件、主题和样式(Style)配置文件、图片资源、菜单布局文件、自动生成的 R.java、apk(可执行的安装包)文件、库文件等。

上述的一系列程序框架不用开发者自己手动管理,在每次建立好一个 Android 项目后,ADT 都会帮我们把这套框架生成好,而开发者需要做的事情就是在框架的每个条目下加入新的内容即可。当然,不用开发者去修改所有的内容,一些是需要修改的,而另一些是系统自动生成、不用修改的。以 2.2.1 节的 Android_Demo01 项目为例,一个一般的 Android 应用程序框架如图 2-18 所示。

图 2-18 所示的 Android 应用程序框架中包含的一些文件和文件夹,具有如下作用。

(1) src/:Java 源代码的存放位置,其和普通 Java 工程中的 src 目录是一样的。在上面项目工程中的 MainActivity.java 就在这个文件夹下。

（2）gen/：存放系统自动生成的配置文件，开发者不要更改。其中的 R.java 是资源索引文件，将 XML 文件中的资源映射为一个资源 ID，供 Java 代码使用。Android 开发工具会自动根据放入 res 目录的 xml 界面文件、图标与常量，同步更新修改 R.java 文件。如果不能生成，则说明 XML 文件中含有错误。

（3）assets/：也可以存放资源，但不会生成资源 ID，需要通过 AssetManager 以二进制流的形式访问。

（4）bin/：存放应用程序编译后生成的可执行文件(.apk)。应用程序中用到的/res/drawable 和/res/raw 下的资源会被打包进 apk 文件。

（5）libs/：存放应用程序所使用的 JAR 包，可以把使用到的第三方 JAR 包放在这。

（6）res/：存放应用程序用到的所有资源，包括界面布局、主题样式、字符串、图片、多媒体资源等。但它和 assets 目录最大的区别在于，res 目录下的资源文件会在 gen 目录下的 R.java 文件中产生以资源文件名命名的静态属性。

该目录下还包括一系列的文件目录，有如下含义。

① drawable-hdpi、drawable-ldpi、drawable-mdpi 目录：分别用于存放高、低、中分辨率的图片或 selector(背景选择器)等，主要是 Android 考虑到为了让图片资源适应各种不同屏幕的分辨率，应用程序会自动根据手机分辨率选择对应的图片资源。

② layout/：存放界面布局的 XML 文件，Java 代码中使用 R.layout.xxx 获得。上面项目工程中的布局文件 activity_main.xml 就在这个文件夹下。

③ menu/：存放选项菜单的 XML 配置文件。

④ raw/：图 2-17 所示的 Android 应用程序框架中并没有包含 raw/文件夹，raw/需要用户手动建立，用以存放多媒体资源，在 Java 代码中可以使用 R.raw.xxx 引用资源。

⑤ values/：存放字符串、颜色资源、尺寸资源的 XML 文件。

⑥ values-v11：定义的主题样式供 API 11 及以上(3.x)的设备使用。

⑦ values-v14：定义的主题样式供 API 14 及以上(4.x)的设备使用。

（7）AndroidManifest.xml：应用级的配置文件，配置一些与应用程序有关的重要信息，包括主包名、权限、程序组件等。这个文件列出了应用程序所提供的功能，在这个文件中，可以指定应用程序使用到的服务(如电话服务、互联网服务、短信服务、GPS 服务等)。另外，新添加一个 Activity 的时候，也需要在这个文件中进行相应的配置，只有配置好后，才能调用此 Activity。

（8）project.properties 文件：项目环境信息，一般不需要修改此文件。

2.2.3　案例程序分析

在 2.2.1 节中我们创建了一个 Android 项目 Android_Demo01，实现了在界面上显示一行文本信息"Hello world!"。下面将对这个项目中生成的代码及主要配置文件进行分析讲解，了解一个简单的 Android 应用程序的原理。

Android_Demo01 项目中主要包括一个布局文件(activity_main.xml)，用于设计用户界面；一个 Activity 组件(MainActivity 类)，用于实现用户界面交互功能；一个配置文件 AndroidManifest.xml。

1. Activity 组件 MainActivity 类

代码如下。

```java
package com.example.android_demo01;

import android.app.Activity;
import android.os.Bundle;
import android.view.Menu;
import android.view.MenuItem;

public class MainActivity extends Activity {

    @Override
    protected void onCreate(Bundle savedInstanceState) {
        super.onCreate(savedInstanceState);
        setContentView(R.layout.activity_main);
    }

    @Override
    public boolean onCreateOptionsMenu(Menu menu) {
        //Inflate the menu; this adds items to the action bar if it is present.
        getMenuInflater().inflate(R.menu.main, menu);
        return true;
    }

    @Override
    public boolean onOptionsItemSelected(MenuItem item) {
        //Handle action bar item clicks here. The action bar will
        //automatically handle clicks on the Home/Up button, so long
        //as you specify a parent activity in AndroidManifest.xml.
        int id=item.getItemId();
        if (id==R.id.action_settings) {
            return true;
        }
        return super.onOptionsItemSelected(item);
    }
}
```

上面代码中的 MainActivity 类继承于 Activity 类，Activity 类是专门负责控制视图 View 与用户进行交互的活动类，可以理解为一个与用户交互的界面。当 MainActivity 启动时会调用该类的 onCreate 方法，从上面代码可以看出，在 onCreate 方法中，首先调用了父类的 onCreate 方法，然后使用 setContentView 方法可以加载布局文件 activity_ main。onCreateOptionsMenu 方法可以初始化 Activity 中的菜单，即加载菜单布局文件 main.xml；当用户单击菜单项时，会调用 onOptionsItemSelected 方法，在此方法中可以处理用户的单击事件。

2. 布局文件 activity_main.xml

代码如下。

```xml
<RelativeLayout xmlns:android="http://schemas.android.com/apk/res/android"
    xmlns:tools="http://schemas.android.com/tools"
    android:layout_width="match_parent"
    android:layout_height="match_parent"
    android:paddingBottom="@dimen/activity_vertical_margin"
    android:paddingLeft="@dimen/activity_horizontal_margin"
    android:paddingRight="@dimen/activity_horizontal_margin"
    android:paddingTop="@dimen/activity_vertical_margin"
    tools:context="com.example.android_demo2_1.MainActivity">

    <TextView
        android:layout_width="wrap_content"
        android:layout_height="wrap_content"
        android:text="@string/hello_world" />

</RelativeLayout>
```

上面代码中＜RelativeLayout＞标签表示界面顶层布局方式是相对布局，其属性 xmlns:android 指定命名空间，顶级元素必须指定命名空间。顶层布局内的＜TextView＞标签表示一个文本框控件，用来显示文本信息，其属性 layout_width 指定该元素的宽度，其中 match_parent 代表填满其父元素，对于顶级元素来说，其父元素就是整个手机屏幕。wrap_content 代表该元素的大小仅包裹其自身内容；属性 android:text 表示文本框控件显示的内容。

3. 应用程序配置文件 AndroidManifest.xml

代码如下。

```xml
<?xml version="1.0" encoding="utf-8"?>
<manifest xmlns:android="http://schemas.android.com/apk/res/android"
    package="com.example.android_demo2_1"
    android:versionCode="1"
    android:versionName="1.0">

    <uses-sdk
        android:minSdkVersion="17"
        android:targetSdkVersion="17" />
```

```xml
<application
    android:allowBackup="true"
    android:icon="@drawable/ic_launcher"
    android:label="@string/app_name"
    android:theme="@style/AppTheme">
    <activity
        android:name=".MainActivity"
        android:label="@string/app_name">
        <intent-filter>
            <action android:name="android.intent.action.MAIN" />
            <category android:name="android.intent.category.LAUNCHER" />
        </intent-filter>
    </activity>
</application>
</manifest>
```

AndroidManifest.xml是每个Android程序中必需的文件,它位于整个项目的根目录,描述了应用程序包中暴露的组件(Activity、Service等),以及它们各自的实现类、各种能被处理的数据和启动位置。AndroidManifest.xml除了能声明程序中的Activity、Service、ContentProviders和Intent Receivers之外,还能指定Permissions(声明权限)和Instrumentation(安全控制和测试)。

上面AndroidManifest.xml文件中各个标签的含义如下。

1) 第一层<Manifest>（属性）

(1) xmlns:android

定义Android命名空间,一般为http://schemas.android.com/apk/res/android,这样使得Android中的各种标准属性能在文件中使用,提供了大部分元素中的数据。

(2) package

指定本应用内Java主程序包的包名,也是一个应用进程的默认名称。

(3) versionCode

用于设备程序识别版本(升级),必须是一个int类型的值,代表程序更新过多少次。

(4) versionName

这个名称是给用户看的,可以将应用程序版本号设置为1.1版,后续更新版本设置为1.2、2.0版本等。

2) 第二层<Application>属性

一个AndroidManifest.xml中必须含有一个<Application>标签,这个标签声明了每一个应用程序的组件及其属性(如icon、label、permission等)。

(1) android:description/android:label

此两个属性都是为许可提供的,均为字符串资源,当用户去看许可列表(android:label)或者某个许可的详细信息(android:description)时,这些字符串资源就可以显示给用户。

（2）android：icon

声明整个应用程序的图标。

（3）android：name

为应用程序所实现的 Application 子类的全名。当应用程序进程开始时,该类在所有应用程序组件之前被实例化。

（4）android：theme

一个资源的风格,它定义了一个默认的主题风格给所有的 Activity,类似 style。

3) 第三层<Activity>属性

（1）android：label

Activity 的名称。

（2）android：name

表示该 Activity 显示的标题。

4) 第四层<intent-filter>属性

<intent-filter>内会设定的内容包括 action、data 与 category 三种。

（1）action 属性

只有 android：name 这个属性。常见的 android：name 值为 android.intent.action.MAIN,表明此 Activity 是作为应用程序的入口。

（2）category 属性

常见的 android：name 值为 android.intent.category.LAUNCHER,决定应用程序是否显示在程序列表里。

5) 第二层<uses-sdk />属性

描述应用所需的 API Level,即版本,在此属性中可以指定支持的最小版本、目标版本以及最大版本。

2.2.4　Android 的应用程序组件

Android 的一个核心特性就是一个应用程序可作为其他程序中的元素(那些允许这样的应用程序提供)。例如,如果程序需要用某些控件来编辑一些图片,另一个程序已经开发出了此项功能,且可供其他程序使用,就可以直接调用那个程序的功能,而不是自己再开发一个。为了实现这样的功能,Android 系统必须能够在需要应用程序中任何一部分时启动它的进程,并且实例化那部分的 Java 对象。为此,不像大多数其他系统中的程序,Android 程序不是只有单一的进入点(如没有 main 方法),而是它们拥有系统实例化和运行必需的组件。

在 2.2.1 节和 2.2.2 节,我们学习了如何创建一个 Android 应用程序,并了解了 Android 的应用程序结构,接下来要学习 Android 应用程序中的 4 个重要组成部分,也就是"应用组件",即 Android 的 4 大应用程序组件。

1. Activity——活动

Activity 为用户提供了一个可视的用户界面。例如,一个短信程序可能有一个

activity 用来显示可以发送信息的联系人,第二个 activity 用来向选中的联系人写消息,其他的 activity 用来查看以前的消息,或者更改设置。虽然应用程序中的各个 activity 所提供的用户界面聚合性很强,但是每个 activity 都独立于其他的 activity,每一个实例化的 activity 都是 Activity 的子类。

一个应用程序由一个或多个 activity 组成,需要多少个取决于应用程序和它的设计。典型的,当应用程序启动的时候,activity 中的一个要首先显示给用户。从一个 activity 移动到另一个,是通过当前的 activity 启动下一个来完成的。

每个 activity 都有一个默认的窗口。一般的情况是,这个窗口填满屏幕,但是它也可以小于屏幕和浮动在其他窗口的上面。窗口中的可视内容被一系列层次的视图(view,派生自 View 类的对象)提供。每个视图都控制了窗口中的一个矩形区域。父视图包含和组织子视图的布局。Android 提供了很多现成的视图以供使用,包括按钮、文本框、滚动条、菜单项、复选框等。整个视图层次通过 Activity.setContentView() 方法放到 activity 的窗口上。

2. service——服务

service 没有用户界面,但它会在后台一直运行。例如,service 可能在用户处理其他事情的时候播放背景音乐,或从网络上获取数据。每个 service 都扩展自 Serivce 类。

多媒体播放器播放音乐是应用 service 的一个非常好的例子。多媒体播放器程序可能含有一个或多个 activity,用户通过这些 activity 选择并播放音乐。然而,音乐回放并不需要一个 activity 来处理,因为用户可能会希望音乐一直播放下去,即使退出了播放器去执行其他程序也不停止。为了让音乐一直播放,多媒体播放器 activity 可能会启动一个 service 在后台播放音乐。Android 系统会使音乐回放 service 一直运行,即使在启动这个 service 的 activity 退出之后。

应用程序可以连接到一个正在运行中的 service。当连接到一个 service 后,可以使用这个 service 向外暴露的接口与这个 service 进行通信。对于上面提到的播放音乐的 service,这个接口可能允许用户暂停、停止或重新播放音乐。与 activity 以及其他组件一样,service 同样运行在应用程序进程的主线程中。所以它们不能阻塞其他组件或用户界面,通常需要为这些 service 派生一个线程执行耗时的任务。

3. broadcast Receiver——广播接收器

broadcast receiver 不执行任何任务,仅仅是接收并响应广播通知的一类组件。大部分广播通知是由系统产生的,例如改变时区、电池电量低、用户选择了一幅图片或者用户改变了语言首选项。应用程序同样也可以发送广播通知,例如通知其他应用程序某些数据已经被下载到设备上可以使用。

一个应用程序可以包含任意数量的 broadcast receiver 来响应它认为很重要的通知。所有的 broadcast receiver 都扩展自 BroadcastReceiver 类。

broadcast receiver 不包含任何用户界面。然而它们可以启动一个 activity 以响应接收到的信息,或者通过 NotificationManager 通知用户。可以通过多种方式使用户知道有新的通知产生,如闪动背景灯、振动设备、发出声音等。通常程序会在状态栏上放置一个持久的图标,用户可以打开这个图标并读取通知信息。

4. content provider——内容提供者

应用程序可以通过 content provider 访问其他应用程序的一些私有数据,这是 Android 提供的一种标准的共享数据的机制。共享的数据可以存储在文件系统中、SQLite 数据库中或其他的一些媒体中。content provider 扩展自 ContentProvider 类,通过实现此类的一组标准的接口可以使其他应用程序存取由它控制的数据。然而应用程序并不会直接调用 ContentProvider 中的方法,而是通过类 ContentResolver。ContentResolver 能够与任何一个 ContentProvider 通信,它与 ContentProvider 合作管理进程间的通信。

当 Android 系统收到一个需要某个组件进行处理的请求的时候,Android 会确保处理此请求的组件的宿主进程是否已经在运行,如果没有,则立即启动这个进程。当请求的组件的宿主进程已经在运行,它会继续查看请求的组件是否可以使用,如果不能立即使用,它会创建一个请求的组件的实例来响应请求。

习 题 2

1. 说明 adb 及其作用。
2. 什么是 DDMS? 它在 Android 的开发中有什么功能?
3. 说明 AVD 在 Eclipse 开发环境中的创建方法。
4. 如何创建及运行一个 Android 项目?
5. 说明 Android 项目中 AndroidManifest.xml 文件的作用。

第 3 章

Android UI 编程

第 2 章讲述了一个简单的 Android 应用程序项目的创建及运行,示例中的 Activity 使用了 XML 布局文件来描述界面如何去组织、定义。大部分情况下,Android 中的一个屏幕就是一个 Activity,通过 Activity,Android 应用可以将信息显示给用户,用户也可以与 Android 应用程序进行交互,而这些具体是如何做到的?本章所讲的内容就可以回答这个问题。

本章介绍 Android UI(User Interface,用户界面)开发中很基础、很常用的内容,这些内容虽然简单,但是会应用在几乎所有的 Android 应用程序设计中,只有很好地掌握它们,才能逐步深入学习 Android UI 开发的高级知识。

3.1 Android 常用 UI 组件

3.1.1 View 和 ViewGroup

Android 中所有的 UI 元素都是使用 View 和 ViewGroup 对象建立的,View 是一个可以将一些信息绘制在屏幕上并与用户产生交互的对象,而 ViewGroup 是一个包含多个 View 和 ViewGroup 的容器,用来定义 UI 布局。

Android 提供了一系列的 View 和 ViewGroup 的子类,开发者可以灵活地组合使用它们来完成界面布局、界面元素绘制和用户交互等工作。同时,开发者还可以选择性地继承一些系统提供的 View,来自定义 View,把自己定义的界面元素显示给用户。

Android 使用 View 类作为界面开发的超类,所有的界面开发都与 View 有关。多个 View 是一个 ViewGroup,但 ViewGroup 本身继承自 View,所以,Android 界面由 View 和 ViewGroup 任意组合而成,Android 的界面开发其实就是对 View 及其各种子孙类做操作。

Android 的 UI 开发使用层次模型来完成,一般都是在一个 ViewGroup 中嵌套多层 ViewGroup,每一层中含有任意数目的 View。可以将整个屏幕看作一个 ViewGroup,它同时也是一个 View,而在这个整体的 ViewGroup 之中,又有多个 ViewGroup 和 View,每个 ViewGroup 中又可以有多个子 ViewGroup 和 View,其基本结构如图 3-1 所示。

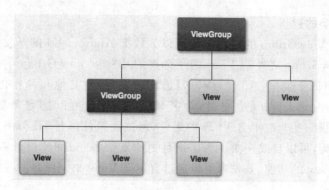

图 3-1　View 和 ViewGroup 基本结构

需要注意的是，虽然 Android 没有规定嵌套的层次深度上限，但是经过大量的实际测试证明，以目前主流的 Android 手机配置，如果一个界面超过 10 层嵌套，它的性能会大幅下降，这个界面的显示会变慢，所以一般不要让嵌套的层次太多。图 3-1 中所示的 View 和 ViewGroup 基本结构中有三层嵌套，它的性能就可以完全得到保证。

由于 View 和 ViewGroup 在 UI 开发中占有很重要的作用，所以有必要讲解一下两者的子孙类继承关系，如图 3-2 所示，图中的空心箭头表示"继承自"的关系。

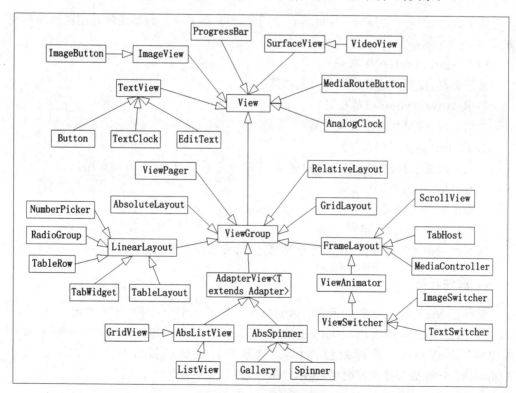

图 3-2　View 的子孙类

如图 3-2 所示，View 的子孙类大体上由 ViewGroup 类和一些单个的界面元素组成，而 ViewGroup 由几个与布局有关的子类继承，Android 的 UI 开发就是组合使用并扩展这些 View 子孙类的过程。

从 View 和 ViewGroup 的继承图中可以大致把 Android UI 的开发划分为两部分，分别是除了 ViewGroup 之外的 View 的子孙类的开发和 ViewGroup 子孙类的开发，但是在 ViewGroup 的子类中，AdapterView（适配器控件）与其他几种相比是有很大区别的，它的内容是由 Adapter（适配器）为其填充的，所以可以把它单独划分为一类。Android 中还提供一些非继承自 View 的组件，但这些组件都与 UI 开发相关，所以也属于 UI 开发的范畴，可以自成一类。为了满足更加复杂多变的 UI 设计需求，Android 还提供了自定义 View 的功能，开发者可以通过自定义 View 来完成各种各样的 UI 设计与显示任务。

3.1.2 Android UI 开发概述

Android 中使用一个叫作 widget 的集合来描述所有的 View 元素，可称其为 View 控件，它包括界面布局、控件和 AdapterView 等。综上所述，可以得到如下所述的 5 种 Android UI 开发分类。

1. 界面布局开发

除 AdapterView 之外的 ViewGroup 的子孙类。以下列出的是几种常用的界面布局方式。

1）LinearLayout（线性布局）
常用布局，按照横或竖的线性排列布局。

2）RelativeLayout（相对布局）
常用布局，按照相对位置来布局。

3）FrameLayout（帧布局）
一块在屏幕上提前预订好的空白区域，可以填充一些 View 元素到里面。

4）TableLayout（表格布局）
以表格的形式布局。

5）AbsoluteLayout（绝对布局）
通过绝对位置坐标布局，但不能做屏幕适配，故不推荐使用。

2. 控件开发

继承自 View 的单个界面元素。Android 自带了几十个控件，常见的有 TextView（文本框）、EditText（编辑框）、Button（按钮）、CheckBox（复选框）、ImageView（图片显示控件）、VideoView（视频播放控件）等。这些控件一般通过各种界面布局方式或 AdapterView 等被安排在界面中，显示给用户，并与用户进行交互。

3. AdapterView 与 Adapter 开发

AdapterView 的子孙类。除了界面布局方式和基本的控件之外，Android 中还提供了多个与适配器相关的控件，这些控件都使用一个适配器来决定该控件显示的内容，它通常是一个列表，其中的数据由适配器提供，而数据源则比较灵活，可以是程序内部数据、本地数据或网络数据等。常用的系统自带的 Adapter 有 BaseAdapter、SimpleAdapter 和 SimpleCursorAdapter 等，除此之外，开发者还可以继承 BaseAdapter 来自定义 Adapter。常见的与适配器相关的控件有 ListView、Spinner、Gallery 和 GridView 等。

4. UI 组件开发

与 UI 相关的 Android 组件。Android 提供了包括之前讲过的 4 大组件在内的一套组件，其中有一些是与 UI 开发相关的，一般将这些与 UI 开发相关的组件称为 UI 组件，它们让开发者在设计 UI 时，多出了一种不使用 View 和 ViewGroup 对象的选择。

UI 组件提供了一套标准化的 UI 布局，开发者只需要简单地指定其中的内容，这些组件即可按照各自的布局格式将信息显示在屏幕上，所以使用它们变得非常方便。这些组件通常不是继承自 View 或 ViewGroup，但是它们仍然是使用 View 的子孙类将信息显示在屏幕上的，所以它们与 View 的关系非常密切。

常用的 UI 组件有 Menu、ActionBar(4.0 的新特性)、Dialog 和 Notification 等，其中 Menu 是 View 的子类，而后三者不是。

5. 自定义 View、图形图像和动画

无论是控件，还是 UI 组件，开发者都可以自定义其中的界面布局样式，通常的做法是继承一个 View 或其子孙类，然后重写一些方法，一般都需要重写 onDraw()方法，该方法用来定义在屏幕上如何进行绘制。

自定义 View 在界面美化、视频图像处理和游戏开发等技术中常被使用。图形图像处理中也大量地应用到了自定义 View。Android 中提供了一些类库，可以实现动画效果。

3.1.3 文本框与编辑框

1. TextView

TextView 是 Android 中最基本的控件，它直接继承自 View，用来向用户显示文本。它有很多子孙类，其中包括之后将要讲到的 EditText 和 Button，还有 DigitalClock、TextClock、CheckedTextView、Chronometer、CheckBox、RadioButton 和 Switch 等。

对于 Android 中的控件，除了在 XML 文件中设置它的属性之外，也可以在 Java 代码中调用它的一系列方法来设置这些属性。所以，这些 XML 属性与 Java 代码中调用的方法是相互对应的。TextView 的常用 XML 属性及在 Java 代码中对应的方法如表 3-1 所示。表 3-1 中列出的属性不包括 android:layout_xxx 系列的属性，因为这些属性是与布局有关的，而与控件本身无关，所有控件都具有这些属性，而且它们的意义也相同，故这里不予列出。

表 3-1 TextView 的常用 XML 属性表

XML 属性	对应的方法	说　明
android:autoText	setKeyListener(KeyListener)	如果设置，即指明这个 TextView 具有一种文本输入法，并会自动纠正一些常见的拼写错误
android:drawableBottom	setCompoundDrawablesWithIntrinsicBounds(int,int,int,int)	文本信息下面显示 Drawable 对象
android:drawableLeft	setCompoundDrawablesWithIntrinsicBounds(int,int,int,int)	文本信息左边显示 Drawable 对象
android:editable		设置其是否可被编辑
android:fontFamily	setTypeface(Typeface)	字体设置
android:hint	setHint(int)	当文本为空时建议它显示的内容
android:gravity	setGravity(int)	指定当文本内容比所在的 TextView 小时，如何放置这些文本
android:inputMethod	setKeyListener(KeyListener)	指定输入法
android:inputType	setRawInputType(int)	指定文本信息的格式，是文字、数字还是时间等
android:lines	setLines(int)	设置 TextView 为多少行高
android:singleLine	setTransformationMethod(TransformationMethod)	设置是一行还是多行
android:text	setText(CharSequence,TextView.BufferType)	最常用的属性，设置其上显示的文本内容
android:textAppearance		设置基本的字体外观
android:textColor	setTextColor(int)	设置字体颜色
android:textSize	setTextSize(int,float)	设置字体大小

除了以上一些与 XML 属性对应的方法之外，TextView 还有很多不与 XML 属性对应的方法，一些常用的方法如表 3-2 所示。

表 3-2 TextView 的常用方法

方法名称	方法前缀	说　明
findViewsWithText（ArrayList＜View＞ outViews，CharSequence searched，int flags）	public void	找到含有给定文本的 View，将其放入一个 ArrayList＜View＞中
getLineCount()	public int	返回行数
getText()	public CharSequence	返回其中显示的文本
length()	public int	返回其中显示的文本的字符数
onTouchEvent(MotionEvent event)	public boolean	实现这个方法去处理在这个 TextView 之上的屏幕触控事件

TextView 不仅能够显示简单的文本，也可以显示复杂的文本，即"富文本"。可以使用 setText()方法设置 TextView 要显示的文本，但是可以指定被显示的文本为一段 HTML 内容，而这段 HTML 内容则完全使用 HTML 的语法，使得 TextView 会显示这段 HTML 定义的界面，这样就大大地扩展了 TextView 可显示的内容，例如显示一个超链接、一张图片，甚至是一个网页。

当写好 HTML 内容之后，需要调用 Html.fromHtml()方法将 HTML 内容转化为 HTML 文本，然后使用 TextView 的 setText()方法设置显示内容即可。如果一个含有"富文本"的 TextView 中包含超链接，默认地，当用户点击它时，只会作为对整个控件的点击事件而响应，并不会响应对超链接的点击，这是因为 Android 无法获得 TextView 的内部元素焦点，需要另外使用 TextView 的 setMovementMethod(LinkMovementMethod.getInstance())方法获得控件内部元素焦点。实现在 TextView 中显示超链接并响应用户点击事件的关键代码参考如下。

```java
//根据一个控件的 Id 获取到该控件的对象
TextView tv=(TextView) findViewById(R.id.textView1);
//为了明显地看到 TextView 的范围，设置其背景色为灰色
tv.setBackgroundColor(getResources().getColor(
        android.R.color.darker_gray));
StringBuilder sb=new StringBuilder();
//放置一个字符串
sb.append("<font color='Red'>All in one TextView!</font><br>");
//放置一个超链接,指向安卓越的官网,点击后会用 WebView 控件显示该网站
sb.append("<font color='Blue'><big><u><a href='http://www.anjoyo.com'>
        Show Anjoyo Home Page</a></u></big></font>");
//将 HTML 的界面设置为 TextView 的内容
tv.setText(Html.fromHtml(sb.toString()));
//使 TextView 内部的超链接可以响应用户的点击
tv.setMovementMethod(LinkMovementMethod.getInstance());

//TextView 被点击的事件处理
tv.setOnClickListener(new View.OnClickListener() {
    @Override
    public void onClick(View v) {
        //使用一个 Toast 控件显示提示信息
        Toast.makeText(getApplicationContext(), "TextView 被点击",
            Toast.LENGTH_LONG).show();
    }
});
```

上面代码的运行效果如图 3-3 所示，其中图 3-3(a)是 TextView 的外观；图 3-3(b)是将语句"tv.setMovementMethod(LinkMovementMethod.getInstance());"代码注释之后，单击超链接的效果，可以看出 TextView 不能获得焦点，自然也就不能打开网站；图 3-3(c)是取消上面语句的注释后，单击 TextView 中非超链接位置的效果；图 3-3(d)是

单击超链接时,获得焦点的效果;图 3-3(e)是单击超链接之后,Android 使用 WebView 控件打开超链接并显示网站的效果;图 3-3(f)为成功显示所指定的网站之后的效果。

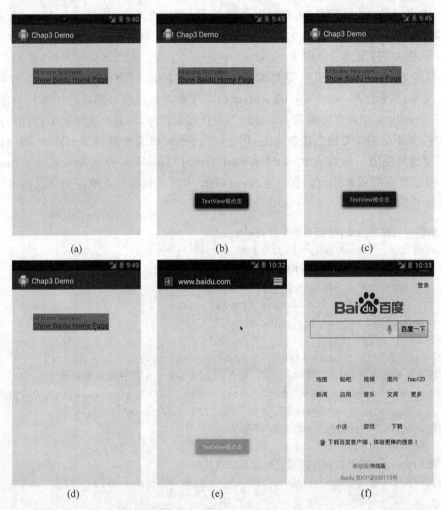

图 3-3　TextView 显示带超链接的富文本

TextView 除了可以显示带超链接的文本之外,还可以在其内部显示图片,实现该功能的关键代码参考如下。

```
//在 TextView 中显示图片
private void showPicInTextView() {
    TextView tv= (TextView) findViewById(R.id.textView1);
    String html="<img src='ic_launcher'/>";
    CharSequence charSequence=Html.fromHtml(html, new ImageGetter() {
        //实现 ImageGetter 接口的 getDrawable()方法,得到图片资源的 Drawable 对象,
        //填充给 TextView
        @Override
        public Drawable getDrawable(String source) {
```

```
                Drawable drawable=getResources().getDrawable(
                        getResourceId(source));
                drawable.setBounds(0, 0, drawable.getIntrinsicWidth(),
                        drawable.getIntrinsicHeight());
                return drawable;
            }
        }, null);
        tv.setText(charSequence);
        tv.setMovementMethod(LinkMovementMethod.getInstance());
        tv.setOnClickListener(new View.OnClickListener() {
            @Override
            public void onClick(View v) {
                //使用一个 Toast 控件显示提示信息
                Toast.makeText(getApplicationContext(), "TextView被点击",
                        Toast.LENGTH_LONG).show();
            }
        });
    }

    //由于无法直接使用文件名来引用图像资源,故利用反射技术获得图像资源 ID
    private int getResourceId(String imageName) {
        int id=0;
        try {
            Field field=R.drawable.class.getField(imageName);
            id=Integer.parseInt(field.get(null).toString());
        } catch (Exception e) {
            e.printStackTrace();
        }
        return id;
    }
```

上面代码的运行效果如图 3-4 所示。

2. EditText

EditText 继承自 TextView,专门用来获取用户输入的文本信息,可以说 EditText 就是一个可编辑的 TextView。在使用 EditText 前,往往先指定 EditText 允许输入的最大行数、输入内容的类型、字体外观等信息。在代码中调用 EditText 的 getText()方法,将得到 Editable 的对象,再调用 toString()方法即可得到用户在 EditText 中输入的内容。

由于 EditText 继承自 TextView,所以 EditText 的

图 3-4　TextView 显示图片

属性与 TextView 的基本相同,但是它还有几个比较重要的属性,需要特别说明如下。

1) android:imeOptions

设置软键盘的 Enter 键。

2) android:autoText

自动拼写帮助,这里单独设置是没有效果的,需要其他输入法辅助。

3) android:inputType

设置文本的类型,用于帮助输入法显示合适的键盘类型,这是 EditText 很重要的一个属性,可设置的文本类型及说明如表 3-3 所示。

表 3-3 EditText 的 inputType 属性可选值

inputType 属性可选值	说明
text	一般文本
textCapCharacters	所有字母大写
textCapWords	每个单词的首字母大写
textCapSentences	仅第一个字母大写
textAutoCorrect	自动更正
textAutoComplete	自动补全
textMultiLine	多行输入
textNoSuggestions	不加提示
textEmailAddress	输入内容为电子邮件地址格式
textEmailSubject	邮件主题格式
textShortMessage	短消息格式,会打开一个表情列表
textLongMessage	长消息格式
textPersonName	人名格式
textPostalAddress	地址格式
textPassword	密码格式
textVisiblePassword	可见密码格式
textWebEditText	输入内容作为网页表单内容
textPhonetic	拼音输入
numberSigned	带符号、数字的格式
numberDecimal	带小数点的浮点格式
phone	电话号码格式
datetime	时间日期格式

4) android:ems

设置 TextView 的宽度为 N(N 为一个整数值)个字符,当设置该属性后,控件显示的长度就为 N 个字符的长度,超出的部分不显示。

5) android:maxLength

最大可输入的字符数。

6) android:password

如果设置 password 属性为 true,则以点(·)显示文本。

3.1.4　按钮与图片视图

从图 3-2 中可以看到,Button 同样继承自 TextView。Button 本身还有几个子孙类,包括 CheckBox、RadioButton、Switch 和 ToggleButton 等。按钮在 Android UI 开发中用得很多,常用来响应用户的点击。

使用 Button 时,首先在 XML 文件中定义,一般要指定其上显示的文本信息,即按钮名称,然后就可以在代码中定义单击它所完成的事件,这样用户在单击它时,就会自动运行点击事件的代码段。

在编写处理点击事件的代码时,有如下两种常用的方法。

一种是先定义一个 OnClickListener 对象,同时实现它的 onClick(View)方法,在该方法中编写按钮被单击后所执行的代码,然后将这个 OnClickListener 对象传递给 Button 对象,参考示例代码如下。

```
Button bt=(Button) findViewById(R.id.button1);
OnClickListener listener=new OnClickListener() {
    @Override
    public void onClick(View v) {
        /*在这里编写按钮被单击后执行的代码*/
    }
};
bt.setOnClickListener (listener);
```

另外一种方法比较简单也更加常用,是直接声明一个匿名类对象并实现 onClick(View)方法,然后将这个对象直接传递给 Button 对象,同样地,也在 onClick(View)方法中处理按钮点击事件,这种方法的参考示例代码如下。

```
Button bt=(Button) findViewById(R.id.button1);
bt.setOnClickListener(new View.OnClickListener() {
    @Override
    public void onClick(View v) {
        /*处理按钮单击事件*/
    }
});
```

Button 同样继承自 TextView,它在 TextView 的属性基础上,也有几个比较重要的属性。

1. android:visibility

此属性的意思是此 Button 是否显示,有三个属性值:visible 表示显示;invisible 表示显示黑背景条；gone 表示不显示。在 Java 代码中，可以设置其显示与否：setVisibility(View. GONE)为不显示；setVisibility(View. VISIBLE)为显示。

2．android：clickable

设置能否被单击。

3．android：focusable

设置能否获得焦点。

4．android：alpha

设置透明度，其值为 0~1 之间的数值，0 为透明，1 为不透明。

5．android：longClickable

设置能否被长按。

3.1.5　案例 ImageView 和 ImageButton

ImageView 控件可以显示图片，ImageButton 控件继承于 ImageView，是一个图片按钮。下面将通过一个案例来演示说明 ImageView 和 ImageButton 的基本使用方法。

1．案例功能描述

案例将实现在界面上显示两行文本信息、一张图片和一个图片按钮，当单击图片按钮后会显示一行文本信息。

2．案例程序结构

案例中包括一个布局文件（image_widget. xml），用于设计用户界面；一个 Activity 组件（MainActivity 类），用于实现用户界面交互功能。

3．案例的实现步骤和思路

第一步，创建一个 Android 项目。

第二步，在 res 目录下的 layout 子目录中创建一个新的布局文件 image_widget. xml，最外层容器为 LinearLayout 布局，其宽和高属性都是充满父容器（match_parent），控件纵向摆放，外层容器之内具体包括以下内容。

（1）创建第一个内嵌布局为线性布局，控件横向摆放，宽度为充满父容器，高度为包裹内容（wrap_content）；

（2）在第一个内嵌布局中创建第一个控件 TextView，宽和高都是包裹内容，靠左排版，显示文本信息"图片"；

（3）在第一个内嵌布局中创建第二个控件 ImageView，宽和高都是包裹内容，靠右排版，显示图片信息 ic_launcher，其 id 为 iv；

(4) 创建第二个内嵌布局为线性布局,控件横向摆放,宽度为充满父容器,高度为包裹内容;

(5) 在第二个内嵌布局中创建第一个控件 TextView,宽和高都是包裹内容,靠左排版,显示文本信息"图片按钮";

(6) 在第二个内嵌布局中创建第二个控件 ImageButton,宽和高都是包裹内容,靠右排版,id 为 ib;

(7) 在最外面的线性布局中创建控件 TextView,宽和高都是包裹内容,其 id 为 tv。

第三步,编写 MainActivity.java 文件,复写生命周期方法 onCreate,具体包括以下步骤。

(1) 先调用父类的 onCreate 方法;

(2) 使用 setContentView 方法加载刚创建的布局文件 image_widget.xml;

(3) 根据界面控件 id 使用 findViewById 方法,获取界面控件 ImageButton 并赋值给局部变量 imageBt;

(4) 根据界面控件 id 使用 findViewById 方法,获取界面控件 TextView 并赋值给 final 型的局部变量 tv,使用 final 关键字是因为在下面的匿名类对象中要使用到这个局部变量;

(5) 为界面控件 ImageButton 对象 imageBt 设置背景图片,图片资源为 ic_launcher;

(6) 为界面控件 ImageButton 设置监听器,使用匿名内部类来创建监听器对象;

(7) 复写监听器对象的 onClick 方法,完成单击后的事件处理;

(8) 在 onClick 方法中给获取到的界面控件 TextView 对象 tv 设置文本信息。

4. 案例参考代码

(1) 布局文件 image_widget.xml 代码如下。

```xml
<?xml version="1.0" encoding="utf-8"?>
<!--最外面的布局文件为线性布局,控件纵向摆放 -->
<LinearLayout xmlns:android="http://schemas.android.com/apk/res/android"
    xmlns:tools="http://schemas.android.com/tools"
    android:layout_width="match_parent"
    android:layout_height="match_parent"
    android:orientation="vertical"
    tools:context=".ImageWidgetActivity">
    <!--第一个内嵌布局为线性布局,控件横向摆放 -->
    <LinearLayout
        android:layout_width="match_parent"
        android:layout_height="wrap_content"
        android:orientation="horizontal">
        <!--第一个内嵌布局中的第一个控件为 TextView,用于显示文本信息 -->
        <TextView
            android:layout_width="wrap_content"
            android:layout_height="wrap_content"
```

```xml
            android:layout_gravity="left"
            android:text="图片" />
        <!--第一个内嵌布局中的第二个控件为ImageView,用于显示图片 -->
        <ImageView
            android:id="@+id/iv"
            android:layout_width="wrap_content"
            android:layout_height="wrap_content"
            android:layout_gravity="right"
            android:src="@drawable/ic_launcher" />
    </LinearLayout>
    <!--第二个内嵌布局为线性布局,控件横向摆放 -->
    <LinearLayout
        android:layout_width="match_parent"
        android:layout_height="wrap_content"
        android:orientation="horizontal">
        <!--第二个内嵌布局中的第一个控件为TextView,用于显示文本信息 -->
        <TextView
            android:layout_width="wrap_content"
            android:layout_height="wrap_content"
            android:layout_gravity="left"
            android:text="图片按钮" />
        <!--第二个内嵌布局中的第二个控件为ImageView,用于显示图片按钮 -->
        <ImageButton
            android:id="@+id/ib"
            android:layout_width="wrap_content"
            android:layout_height="wrap_content"
            android:layout_gravity="right" />
    </LinearLayout>
    <!--最外面的线性布局内嵌控件为TextView,用于显示文本信息 -->
    <TextView
        android:id="@+id/tv"
        android:layout_width="wrap_content"
        android:layout_height="wrap_content" />
</LinearLayout>
```

(2) Activity 组件 MainActivity 类代码如下：

```java
package com.example.android_demo3_1
import android.app.Activity;
import android.os.Bundle;
import android.view.View;
import android.widget.ImageButton;
import android.widget.TextView;


/* 主Activity为控制类,用于控制界面的显示 */
```

```java
public class MainActivity extends Activity {
    /*Activity组件的生命周期方法,在组件创建时调用一般用于初始化信息*/
    @Override
    protected void onCreate(Bundle savedInstanceState) {
        //调用父类方法,完成系统工作
        super.onCreate(savedInstanceState);
        //加载界面布局文件
        setContentView(R.layout.image_widget);
        //根据控件id获取界面上的ImageButton控件
        ImageButton imageBt=(ImageButton) findViewById(R.id.ib);
        //根据控件id获取界面上的TextView控件
        final TextView tv=(TextView) findViewById(R.id.tv);
        //给获取到的界面控件ImageButton设置背景图片
        imageBt.setBackgroundResource(R.drawable.ic_launcher);
        //给获取到的界面控件ImageButton设置监听器,使用匿名内部类来创建监听器对象
        imageBt.setOnClickListener(new View.OnClickListener() {
            /*复写监听器对象的onClick方法,完成点击后的事件处理,参数为被点击的控
              件*/
            @Override
            public void onClick(View v) {
                //给获取到的界面控件TextView设置文本信息
                tv.setText("ImageButton Clicked");
            }
        });
    }
}
```

5. 案例运行效果

案例运行效果如图3-5所示。

6. ImageView 和 ImageButton 的 scaleType 属性

ImageView、ImageButton等图片控件都有一个属性,叫作android:scaleType,它指定图片如何放大、缩小或移动来匹配ImageView的大小,其可选值8种,分别介绍如下。

(1) ImageView.ScaleType.CENTER|android:scaleType="center"

按图片的原来大小居中显示,当图片长和宽超过View的长和宽时,则截取部分图片并居中显示。

(2) ImageView.ScaleType.CENTER_CROP|android:scaleType="centerCrop"

按比例扩大图片并居中显示,使得图片长(宽)等于或大于View的长(宽)。

(3) ImageView.ScaleType.CENTER_INSIDE|android:scaleType="centerInside"

将图片的内容完整居中显示,通过按比例缩小原来的大小使得图片长(宽)等于或小于View的长(宽)。

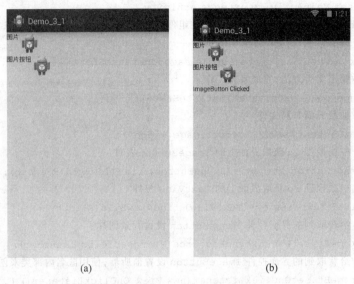

图 3-5 ImageView 和 ImageButton

(4) ImageView.ScaleType.FIT_CENTER|android:scaleType="fitCenter"
把图片按比例扩大或缩小到 View 的宽度,居中显示。
(5) ImageView.ScaleType.FIT_END|android:scaleType="fitEnd"
把图片按比例扩大或缩小到 View 的宽度,显示在 View 的下部分位置。
(6) ImageView.ScaleType.FIT_START|android:scaleType="fitStart"
把图片按比例扩大或缩小到 View 的宽度,显示在 View 的上部分位置。
(7) ImageView.ScaleType.FIT_XY|android:scaleType="fitXY"
把图片按照指定的大小在 View 中显示。
(8) ImageView.ScaleType.MATRIX|android:scaleType="matrix"
用矩阵来绘制,当绘制时使用图片矩阵缩放。图片矩阵可用 setImageMatrix(Matrix)进行设定。

3.1.6 案例 CheckBox、RadioButton 和 ToggleButton

CheckBox、RadioButton 和 ToggleButton 是 Button 的子类,都是与处理用户单击事件有关的控件。其中,CheckBox 是复选框,RadioButton 是单选按钮,而 ToggleButton 是开关按钮。

CheckBox 的使用较简单,它只有选中或未选中两种状态,用户可以通过手指在屏幕上单击来选中或取消选中,而开发者则可以在 Java 代码中通过调用 setOnCheckedChangeListener()方法来处理用户的选中和取消选中事件。

RadioButton 需要使用一个 RadioGroup 来组织多个 RadioButton,在同一个 RadioGroup 中,一次只能选中一个 RadioButton。

ToggleButton 是一种具备两种状态的按钮,常用于表示开/关的场景中,它不同于 Button,特点是可以被按中和不按中的状态,而且在按中时跟未按中时分别可以显示不同的文本。

下面通过一个案例来演示说明 CheckBox、RadioButton 和 ToggleButton 的基本使用方法。

1. 案例功能描述

案例将实现在界面上显示三组内容,第一组是一个复选框及两个 TextView,第二组为单选框按钮及两个 TextView,第三组为开关按钮及两个 TextView。

2. 案例程序结构

案例中包括一个布局文件(button_widget.xml),用于设计用户界面;一个 Activity 组件(MainActivity 类),用于实现用户界面交互功能。

3. 案例的实现步骤和思路

(1) 创建 Android 项目。

(2) 在 res 目录下的 layout 子目录中创建新的布局文件 button_widget.xml,最外层容器为 LinearLayout 布局,宽和高都是充满父容器,控件纵向摆放。

① 创建第一个内嵌布局为相对布局,宽和高都为充满父容器,比重为 10;

② 在第一个内嵌布局中创建第一个控件 TextView,宽和高都是包裹内容,显示文本信息"复选框",字号为 22sp,id 为 cb_name_tv;

③ 在第一个内嵌布局中创建第二个控件 TextView,宽和高都是包裹内容,显示文本信息"未被选中",id 为 cb_tv,在控件 cb_name_tv 的下方、父容器顶部,字号为 18sp;

④ 在第一个内嵌布局中创建第三个控件 CheckBox,宽和高都是包裹内容,显示文本信息"复选框",id 为 check_box,与控件 cb_tv 基线对齐,左外边距为 20dp,在控件 cb_tv 的右边,字号为 18sp;

⑤ 创建第二个内嵌布局为相对布局,宽和高都为充满父容器,比重为 10;

⑥ 在第二个内嵌布局中创建第一个控件 TextView,宽和高都是包裹内容,id 为 rb_name_tv,文本显示"单选",字号为 22sp;

⑦ 在第二个内嵌布局中创建第二个控件 TextView,宽和高都是包裹内容,id 为 rb_tv,在控件 rb_name_tv 的下方,上方外边距为 10dp,字号为 18sp;

⑧ 在第二个内嵌布局中创建第三个控件 RadioGroup,宽和高都是充满父容器,id 为 radio_group,在控件 rb_tv 的下方,选中按钮的 id 为 rb1,单选按钮为横向排列;

⑨ 在 RadioGroup 中创建第一个控件 RadioButton,宽和高都是包裹内容,id 为 rb1,文本显示"单选 1",字号为 18sp;

⑩ 在 RadioGroup 中创建第二个控件 RadioButton,宽和高都是包裹内容,id 为 rb2,文本显示"单选 2",字号为 18sp;

⑪ 在 RadioGroup 中创建第三个控件 RadioButton,宽和高都是包裹内容,id 为 rb3,文本显示"单选 3",字号为 18sp;

⑫ 创建第三个内嵌布局为相对布局,宽和高都为充满父容器,比重为 10;

⑬ 在第三个内嵌布局中创建第一个控件 TextView,宽和高都是包裹内容,显示文本信息"开关按钮",字号为 22sp,id 为 tb_name_tv;
⑭ 在第三个内嵌布局中创建第二个控件 TextView,宽和高都是包裹内容,显示文本信息"开",字号为 18sp,id 为 tb_tv,在控件 tb_name_tv 的下方,顶部外边距为 10dp;
⑮ 在第三个内嵌布局中创建第三个控件 ToggleButton,宽和高都是包裹内容,id 为 toggle_button,在控件 tb_tv 的下方,开状态文本为"开",关状态文本为"关"。
(3) 编写 MainActivity 文件,复写生命周期方法 onCreate。
① 先调用父类的 onCreate 方法;
② 使用 setContentView 方法加载刚创建的新的布局文件 ibutton_widget.xml;
③ 根据界面控件 id 使用 findViewById 方法,获取界面控件 CheckBox 并赋值给局部变量 cb;
④ 根据界面控件 id 使用 findViewById 方法,获取界面控件 TextView 并赋值给局部变量 tv;
⑤ 根据界面控件 id 使用 findViewById 方法,获取界面控件 RadioGroup 并赋值给局部变量 rg;
⑥ 根据界面控件 id 使用 findViewById 方法,获取界面控件 TextView 并赋值给局部变量 tv;
⑦ 根据界面控件 id 使用 findViewById 方法,获取界面控件 ToggleButton 并赋值给局部变量 tb;
⑧ 根据界面控件 id 使用 findViewById 方法,获取界面控件 TextView 并赋值给局部变量 tbTv;
⑨ 为界面控件 CheckBox 设置初始选择状态为不选中,即采用语句"cb.setChecked(false);";
⑩ 为界面控件 CheckBox 设置监听器,使用匿名内部类来创建监听器对象,即采用语句"cb.setOnCheckedChangeListener();";
⑪ 复写监听器对象的 onCheckedChanged 方法,完成单选后的事件处理,即设置 cbTv 控件的显示内容;
⑫ 设置控件 rbTv 的文本显示内容,即采用语句"rbTv.setText("当前选中:" +((RadioButton)(findViewById(rg.getCheckedRadioButtonId()))).getText());";
⑬ 为界面控件 RadioGroup 设置监听器,使用匿名内部类创建监听器对象,即采用语句"rg.setOnCheckedChangeListener(new RadioGroup.OnCheckedChangeListener();";
⑭ 复写监听器对象的 onCheckedChanged 方法,完成单选后的事件处理,即设置 rbTv 控件的显示内容,采用语句"rbTv.setText("当前选中:" +((RadioButton)(findViewById(checkedId))).getText());";
⑮ 为界面控件 ToggleButton 设置初始选择状态为不选中,即采用语句"tb.setChecked(false);";
⑯ 为界面控件 ToggleButton 设置监听器,使用匿名内部类来创建监听器对象;
⑰ 复写监听器对象的 onClick 方法,完成单选后的事件处理,设置文本 tbTv 控件的

显示内容。

4. 案例参考代码

(1) 布局文件 button_widget.xml 代码如下。

```xml
<?xml version="1.0" encoding="utf-8"?>
<!--最外面的布局文件为线性布局,控件纵向摆放 -->
<LinearLayout xmlns:android="http://schemas.android.com/apk/res/android"
    xmlns:tools="http://schemas.android.com/tools"
    android:layout_width="match_parent"
    android:layout_height="match_parent"
    android:orientation="vertical"
    tools:context=".ButtonWidgetActivity">
    <!--第一个内嵌布局为相对布局 -->
    <RelativeLayout
        android:layout_width="match_parent"
        android:layout_height="match_parent"
        android:layout_weight="10">
        <!--第一个内嵌布局中的第一个控件为TextView,用于显示文本信息 -->
        <TextView
            android:id="@+id/cb_name_tv"
            android:layout_width="wrap_content"
            android:layout_height="wrap_content"
            android:text="复选框"
            android:textSize="22sp" />
        <!--第一个内嵌布局中的第二个控件为TextView,用于显示文本信息 -->
        <TextView
            android:id="@+id/cb_tv"
            android:layout_width="wrap_content"
            android:layout_height="wrap_content"
            android:layout_below="@+id/cb_name_tv"
            android:layout_marginTop="10dp"
            android:text="未被选中"
            android:textSize="18sp" />
        <!--第一个内嵌布局中的第三个控件为CheckBox,用于显示复选框 -->
        <CheckBox
            android:id="@+id/check_box"
            android:layout_width="wrap_content"
            android:layout_height="wrap_content"
            android:layout_alignBaseline="@+id/cb_tv"
            android:layout_marginLeft="20dp"
            android:layout_toRightOf="@+id/cb_tv"
            android:text="复选框"
            android:textSize="18sp" />
```

```xml
        </RelativeLayout>
        <!--第二个内嵌布局为相对布局 -->
        <RelativeLayout
            android:layout_width="match_parent"
            android:layout_height="match_parent"
            android:layout_weight="10">
            <!--第二个内嵌布局中的第一个控件为 TextView,用于显示文本信息 -->
            <TextView
                android:id="@+id/rb_name_tv"
                android:layout_width="wrap_content"
                android:layout_height="wrap_content"
                android:text="单选"
                android:textSize="22sp" />
            <!--第二个内嵌布局中的第二个控件为 TextView,用于显示文本信息 -->
            <TextView
                android:id="@+id/rb_tv"
                android:layout_width="wrap_content"
                android:layout_height="wrap_content"
                android:layout_below="@+id/rb_name_tv"
                android:layout_marginTop="10dp"
                android:textSize="18sp" />
            <!--第二个内嵌布局中的第三个控件为 RadioGroup,用于包裹一组单选按钮 -->
            <RadioGroup
                android:id="@+id/radio_group"
                android:layout_width="fill_parent"
                android:layout_height="fill_parent"
                android:layout_below="@+id/rb_tv"
                android:checkedButton="@+id/rb1"
                android:orientation="horizontal">
                <!--RadioGroup 中第一个单选按钮,用于显示单选框 -->
                <RadioButton
                    android:id="@+id/rb1"
                    android:layout_width="wrap_content"
                    android:layout_height="wrap_content"
                    android:text="单选 1"
                    android:textSize="18sp">
                </RadioButton>
                <!--RadioGroup 中第二个单选按钮,用于显示单选框 -->
                <RadioButton
                    android:id="@+id/rb2"
                    android:layout_width="wrap_content"
                    android:layout_height="wrap_content"
                    android:text="单选 2"
                    android:textSize="18sp">
```

```xml
        </RadioButton>
        <!--RadioGroup 中第三个单选按钮,用于显示单选框 -->
        <RadioButton
            android:id="@+id/rb3"
            android:layout_width="wrap_content"
            android:layout_height="wrap_content"
            android:text="单选 3"
            android:textSize="18sp">
        </RadioButton>
    </RadioGroup>
</RelativeLayout>
<!--第三个内嵌布局为相对布局 -->
<RelativeLayout
    android:layout_width="match_parent"
    android:layout_height="match_parent"
    android:layout_weight="10">
    <!--第三个内嵌布局中的第一个控件为 TextView,用于显示文本信息 -->
    <TextView
        android:id="@+id/tb_name_tv"
        android:layout_width="wrap_content"
        android:layout_height="wrap_content"
        android:text="开关按钮"
        android:textSize="22sp" />
    <!--第三个内嵌布局中的第二个控件为 TextView,用于显示文本信息 -->
    <TextView
        android:id="@+id/tb_tv"
        android:layout_width="wrap_content"
        android:layout_height="wrap_content"
        android:layout_below="@+id/tb_name_tv"
        android:layout_marginTop="10dp"
        android:text="开"
        android:textSize="18sp" />
    <!--第三个内嵌布局中的第三个控件为 ToggleButton,用于显示开关按钮 -->
    <ToggleButton
        android:id="@+id/toggle_button"
        android:layout_width="wrap_content"
        android:layout_height="wrap_content"
        android:layout_below="@+id/tb_tv"
        android:textOff="单击以打开"
        android:textOn="单击以关闭" />
</RelativeLayout>
</LinearLayout>
```

(2) Activity 组件 MainActivity 类代码如下。

```java
package com.example.android_demo3_2
import android.app.Activity;
import android.os.Bundle;
import android.view.View;
import android.widget.CheckBox;
import android.widget.CompoundButton;
import android.widget.RadioButton;
import android.widget.RadioGroup;
import android.widget.TextView;
import android.widget.ToggleButton;

/*主Activity为控制类,用于控制界面的显示*/
public class MainActivity extends Activity {
    /*Activity组件的生命周期方法,在组件创建时调用一般用于初始化信息*/
    @Override
    protected void onCreate(Bundle savedInstanceState) {
        //调用父类方法,完成系统工作
        super.onCreate(savedInstanceState);
        //加载界面布局文件
        setContentView(R.layout.button_widget);
        //根据控件id获取界面上的CheckBox控件
        CheckBox cb=(CheckBox) findViewById(R.id.check_box);
        //根据控件id获取界面上的TextView控件
        final TextView cbTv=(TextView) findViewById(R.id.cb_tv);
        //根据控件id获取界面上的RadioGroup控件
        RadioGroup rg=(RadioGroup) findViewById(R.id.radio_group);
        //根据控件id获取界面上的TextView控件
        final TextView rbTv=(TextView) findViewById(R.id.rb_tv);
        //根据控件id获取界面上的ToggleButton控件
        final ToggleButton tb=(ToggleButton) findViewById(R.id.toggle_button);
        //根据控件id获取界面上的TextView控件
        final TextView tbTv=(TextView) findViewById(R.id.tb_tv);
        //给CheckBox设置默认选择状态
        cb.setChecked(false);                    /*设置初始状态为未选中*/
        //给CheckBox的单击事件
            cb.setOnCheckedChangeListener(new
                CompoundButton.OnCheckedChangeListener() {
                /*复写监听器对象的onCheckedChanged方法,完成单选后的事件处理,
                  参数为被单击的控件和复选框选择状态*/
                @Override
                public void onCheckedChanged(
                    CompoundButton buttonView,
                    boolean isChecked) {
                    if (isChecked)              /*如果被选中,在这里处理*/
```

```java
                    cbTv.setText("已被选中");
            else                                  /*如果取消选中,在这里处理*/
                    cbTv.setText("未被选中");
        }
    });
    //给TextView控件设置文本信息
    rbTv.setText("当前选中:"
            +((RadioButton) (findViewById(rg
                    .getCheckedRadioButtonId())))
                    .getText());
    //给RadioButton设置单击事件监听器
    rg.setOnCheckedChangeListener(new RadioGroup.OnCheckedChangeListener() {
        /*复写监听器对象的onCheckedChanged方法,
          完成单选后的事件处理,参数为被单击的单选框组单选框选择状态*/
        @Override
        public void onCheckedChanged(RadioGroup group,
                int checkedId) {
            /*传入的checkedId参数表示当前被用户选中的RadioButton的编号,
              它就是R.id.xxx索引到的Id号,也就是保存在R.java中的Id号,
              所以不能直接使用它,而是需要通过它找到相应的RadioButton,
              然后调用getText()得到它上面的文本*/
            rbTv.setText("当前选中:"
                    +((RadioButton) (findViewById(checkedId)))
                            .getText());
        }
    });
    //给开关按钮设置初始状态
    tb.setChecked(false);                        /*设置初始状态为已关闭*/
    //ToggleButton设置监听器
    tb.setOnClickListener(new View.OnClickListener() {
        /*复写监听器对象的onClick方法,完成单击后的事件处理,参数为被单击的
          按钮*/
        @Override
        public void onClick(View v) {
            if (tb.isChecked())
                tbTv.setText("当前状态为:已打开");
            else
                tbTv.setText("当前状态为:已关闭");
        }
    });
}
}
```

5. 案例运行效果

上面代码的运行效果如图 3-6 所示，图 3-6(a)显示的是复选框没被选中、单选按钮选择"单选 1"、开关按钮没有打开的状态，图 3-6(b)显示的是复选框被选中、单选按钮选择"单选 2"、开关按钮打开的状态。

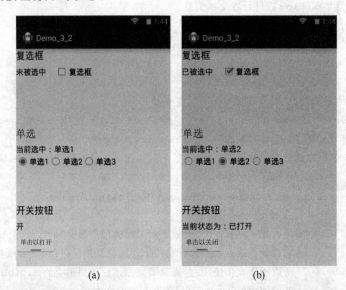

图 3-6　案例 CheckBox、RadioButton 和 ToggleButton 演示效果

3.2　Android 常用布局

前几节讲述了 Android 的一些基本控件。在案例中，我们只是简单地使用了 LinearLayout 布局和 RelativeLayout 布局，LinearLayout 布局只是把控件垂直排列在 Activity 中，这样的布局当然缺乏美感。所以这一节将讲述如何将 UI 控件合理、美观地布局在 Activity 中。

定义 UI 布局的最常用的方法是使用 XML 布局文件，如同 HTML 一样，XML 为布局提供了一种可读的结构。XML 中的每个元素都是 View 或 ViewGroup 的子孙类的对象，可以把每一个 XML 布局文件理解为一棵由 View 和 ViewGroup 的子孙类对象组成的树，树根是一个 ViewGroup 对象，所有的叶结点都是 View 对象，树的分支结点都是 ViewGroup 对象。

开发者有相当多的方法来对视图进行布局，通过使用大量不同种类的 ViewGroup，开发者可以有近乎无穷的方式来构建 View 和 ViewGroup。Android 提供了一些预定义的 ViewGroup 子孙类，常用的有 LinearLayout、RelativeLayout、TableLayout、FrameLayout、GridLayout 和 AbsoluteLayout，下面就对这 6 种布局方式逐一介绍。

3.2.1 线性布局

LinearLayout 线性布局是一种最简单的布局方式,它有垂直和水平两种布局方向,使用 android:orientation="vertical"属性设置可以指定布局方式为垂直,使用 android:orientation= "horizontal"属性设置可以指定布局方式为水平。下面将通过一个案例了解 LinearLayout 这种布局方式。

1. 案例功能描述

案例将实现在界面上显示两组内容,第一组是一个占屏幕 1/3 空间的相对布局,背景为白色;第二组是一个占屏幕 2/3 空间的相对布局,背景为黑色。

2. 案例程序结构

案例中包括一个布局文件(activity_main.xml),用于设计用户界面;一个 Activity 组件(MainActivity 类),用于实现用户界面交互功能。

3. 案例的实现步骤和思路

(1) 创建 Android 项目。

(2) 在 res 目录下的 layout 子目录中创建新的布局文件 activity_main.xml,最外层容器为 LinearLayout 布局,宽和高都是充满父容器,控件纵向摆放。

① 创建第一个内嵌布局为相对布局,宽和高都为充满父容器,比重为 2;

② 创建第二个内嵌布局为相对布局,宽和高都为充满父容器,比重为 1。

(3) 编写 MainActivity 文件,复写生命周期方法 onCreate。

① 先调用父类的 onCreate 方法;

② 使用 setContentView 方法加载刚创建的新的布局文件 activity_main.xml。

4. 案例参考代码

(1) 布局文件 button_widget.xml 代码如下。

```xml
<?xml version="1.0" encoding="utf-8"?>
<!--最外面的布局文件为线性布局,控件纵向摆放 -->
<LinearLayout xmlns:android="http://schemas.android.com/apk/res/android"
    xmlns:tools="http://schemas.android.com/tools"
    android:layout_width="match_parent"
    android:layout_height="match_parent"
    android:orientation="vertical">
    <!--第一个内嵌布局为相对布局 -->
    <RelativeLayout
        android:layout_width="match_parent"
        android:layout_height="match_parent"
        android:layout_weight="2"
```

```xml
        android:background="@android:color/white">
    </RelativeLayout>
    <!--第二个内嵌布局为相对布局 -->
    <RelativeLayout
        android:layout_width="match_parent"
        android:layout_height="match_parent"
        android:layout_weight="1"
        android:background="@android:color/black">
    </RelativeLayout>
</LinearLayout>
```

在布局文件 button_widget.xml 中,最外面的布局文件为线性布局,其属性 orientation 指定子元素排列方式,指定为 vertical 表示子元素垂直排列,每个子元素会占独立的一行,而另一个可选值为 horizontal,它表示子元素水平排列,即每个子元素会占独立的一列。属性 layout_width 指定该元素的宽度,其中 match_parent 代表填满其父元素,对于顶级元素来说,其父元素就是整个手机屏幕;wrap_content 代表该元素的大小仅包裹其自身内容;数字则代表其占相应的像素值。属性 layout_height 指定该元素的高度,可选参数值与 layout_width 的参数意义相同。

LinearLayout 有两个非常相似的属性:android:gravity 和 android:layout_gravity,它们都是用来设置对齐方式的,可选值包括 left(左对齐)、right(右对齐)、top(上对齐)、bottom(下对齐)、center(居中)、center_horizontal(水平居中)和 center_vertical(垂直居中)等,这些值还可以组合使用,中间用|分开。

android:gravity 和 android:layout_gravity 的区别如下。

① android:gravity:用于设置该 View 内部内容的对齐方式。例如一个 Button 的该属性是设置其上的文本在这个 Button 中的位置。

② android:layout_gravity:用于设置该 View 在其父 View 中的对齐方式。例如一个 Button 在一个 LinearLayout 中,可以使用 Button 的该属性设置这个 Button 在这个 LinearLayout 中的位置。

LinearLayout 还有一个非常重要的属性 android:layout_weight,它表示比重的意思,用它就可以实现通过百分比来进行布局的目的,从而使得这样布局之下的 Activity 可以不用专门的屏幕分辨率适配过程,就可以直接在多种不同分辨率的屏幕中正常地显示。

需要注意,百分比布局的属性 android:layout_weight 只在 LinearLayout 中可用,而 RelativeLayout 没有该属性,所以如果想使用百分比布局,则需要保证其父 View 是一个 LinearLayout。

可以通过给 android:layout_weight 属性设置一个数字,来指定一个 LinearLayout 中的各个 View 的百分比空间占用,如上面 XML 文件中的 android:layout_weight="2"。layout_weight 的值与占用比重是相反的,其值设置得越大,表示它占用的比重越小,所以上面 XML 文件中第一个 RelativeLayout 共占用了 1/3 的垂直空间,而第二个 RelativeLayout 占用了 2/3 的垂直空间。

如果将设置了 layout_weight 属性的 View 的 layout_width 或 layout_height 设置为 wrap_content，则对比重的判断会变为正相关，即 layout_weight 值设置得越小，占用的空间越少。但是这种情况下，有时候不会严格地按照比重来显示，如果某个 View 中的内容过多，就会占用过多的空间，只有当各个 View 的内容都少于或等于该 View 的大小时，才会按照定义的比重显示。

（2）Activity 组件 MainActivity 类代码如下：

```
package com.example.android_demo3_3
import android.app.Activity;
import android.os.Bundle;
public class MainActivity extends Activity {
    @Override
    protected void onCreate(Bundle savedInstanceState) {
        super.onCreate(savedInstanceState);
        sctContentView(R.layout.activity_main);
    }
}
```

5. 案例运行效果

LinearLayout 布局案例的运行结果如图 3-7 所示。

3.2.2 相对布局

相对布局中的视图组件是按相互之间的相对位置来确定的，并不是线性布局中的必须按行或按列单个显示，在此布局中的子元素里与位置相关的属性将生效。例如 android：layout_below、android：layout_above 等。注意在指定位置关系时，引用的 ID 必须在引用之前先被定义，否则将出现异常。RelativeLayout 是 Android 布局结构中最灵活的一种布局结构，比较适合一些复杂界面的布局。RelativeLayout 用到的一些重要的属性如表 3-4 所示。

图 3-7　LinearLayout 布局案例运行效果

表 3-4　RelativeLayout 的重要属性

按属性值分类	属 性 名 称	属 性 含 义
属性值为 true 或 false	android：layout_centerHrizontal	水平居中
	android：layout_centerVertical	垂直居中
	android：layout_centerInparent	相对于父元素完全居中
	android：layout_alignParentBottom	贴紧父元素的下边缘

续表

按属性值分类	属性名称	属性含义
属性值为 true 或 false	android:layout_alignParentLeft	贴紧父元素的左边缘
	android:layout_alignParentRight	贴紧父元素的右边缘
	android:layout_alignParentTop	贴紧父元素的上边缘
	android:layout_alignWithParentIfMissing	如果对应的兄弟元素找不到的话就以父元素做参照物
属性值必须为 id 的引用名	android:layout_below	在某元素的下方
	android:layout_above	在某元素的上方
	android:layout_toLeftOf	在某元素的左边
	android:layout_toRightOf	在某元素的右边
	android:layout_alignTop	本元素的上边缘和某元素的上边缘对齐
	android:layout_alignLeft	本元素的左边缘和某元素的左边缘对齐
	android:layout_alignBottom	本元素的下边缘和某元素的下边缘对齐
	android:layout_alignRight	本元素的右边缘和某元素的右边缘对齐
属性值为具体的像素值	android:layout_marginBottom	离某元素底边缘的距离
	android:layout_marginLeft	离某元素左边缘的距离
	android:layout_marginRight	离某元素右边缘的距离
	android:layout_marginTop	离某元素上边缘的距离

下面通过一个案例了解 RelativeLayout 这种布局方式。

1. 案例功能描述

案例将实现在界面上显示一个编辑框 EditText、两个按钮 Button 和一个文本框 TextView。编辑框在界面的左上角，文本框在输入框下面，第一个按钮在文本框的下面，并在编辑框的右边，第二个按钮在第一按钮的右边。在编辑框中输入文本后，单击第一个按钮，可以将输入内容显示在文本框上，单击第二个按钮可以将编辑框内容清空。

2. 案例程序结构

案例中包括一个布局文件（activity_main.xml），用于设计用户界面；一个 Activity 组件（MainActivity 类），用于实现用户界面交互功能。

3. 案例的实现步骤和思路

(1) 创建 Android 项目。

(2) 在 res 目录下的 layout 子目录中创建新的布局文件 activity_main.xml，最外层容器为 RelativeLayout 布局，宽和高都是充满父容器。

① 创建第一个内嵌控件为 EditText，宽 120dp，高为包裹内容，id 为 et，输入类型为 text；

② 创建第二个内嵌控件为 TextView，宽和高为包裹内容，id 为 tv，在控件 et 的下面；

③ 创建第三个内嵌控件为 Button，宽和高为包裹内容，id 为 bt_ok，在控件 et 的右边、控件 tv 的下方，文本显示内容为"确认"；

④ 创建第四个内嵌控件为 Button，宽和高为包裹内容，id 为 bt_clear，在控件 bt_ok 的右边，与控件 bt_ok 顶端对齐，文本显示内容为"清除"。

(3) 编写 MainActivity 文件，复写生命周期方法 onCreate。

① 先调用父类的 onCreate 方法；

② 使用 setContentView 方法加载刚创建的新的布局文件 activity_main.xml；

③ 编写方法 getViewItem，在方法中根据界面控件 id 获取界面的 4 个控件，分别赋值给成员变量 mButtonOK、mButtonCancel、mTextView 和 mEditText；

④ 给两个按钮设置监听器；

⑤ MainActivity 实现 OnClickListener 接口成为单击监听器；

⑥ 实现单击监听器的 onClick 方法；

⑦ 在 onClick 方法中使用 switch 语句，根据不同的按钮 id 区分不同的单击处理；

⑧ 当单击 ok 按钮时获取文件编辑框的输入内容交给 TextView 去显示；

⑨ 当单击 clear 按钮时将编辑框的输入内容清空。

4. 案例参考代码

(1) 布局文件 activity_main.xml 代码如下。

```xml
<?xml version="1.0" encoding="utf-8"?>
<!--最外面的布局文件为相对布局,控件纵向摆放 -->
<RelativeLayout xmlns:android="http://schemas.android.com/apk/res/android"
    xmlns:tools="http://schemas.android.com/tools"
    android:layout_width="match_parent"
    android:layout_height="match_parent">

    <!--定义一个 EditText,用来接收用户输入 -->
    <EditText
        android:id="@+id/et"
        android:layout_width="120dp"
        android:layout_height="wrap_content"
        android:inputType="text" />

    <!--定义一个 TextView,用来显示文本信息 -->
    <TextView
        android:id="@+id/tv"
        android:layout_width="wrap_content"
```

```
        android:layout_height="wrap_content"
        android:layout_below="@+id/et"/>

    <!--定义一个Button,用来响应用户单击 -->
    <Button
        android:id="@+id/bt_ok"
        android:layout_width="wrap_content"
        android:layout_height="wrap_content"
        android:layout_below="@+id/tv"
        android:layout_toRightOf="@+id/et"
        android:text="确认" />
    <!--定义一个Button,用来响应用户单击 -->
    <Button
        android:id="@+id/bt_clear"
        android:layout_width="wrap_content"
        android:layout_height="wrap_content"
        android:layout_alignTop="@+id/bt_ok"
        android:layout_toRightOf="@+id/bt_ok"
        android:layout_marginLeft="25dp"
        android:text="清除" />
</RelativeLayout>
```

在上面的 XML 文件中的最外层布局为<RelativeLayout>标签,说明其中的子元素都以相对位置关系来定义各自的位置。在指定相对位置时,需要明确是以哪一个控件作为基准,所以在 RelativeLayout 中往往需要定义每一个控件的 id。

屏幕左上角的控件是一个 EditText,由于没有指定它的相对位置,所以默认出现在屏幕左上角,而后的是一个 TextView,使用了 android:layout_below="@+id/et"这句代码将其放置在 EditText 的下边,由于没有说明其水平位置,所以默认出现在屏幕左边。

在 TextView 之下是两个 Button,"确定"按钮通过 android:layout_below="@+id/tv"和 android:layout_toRightOf="@+id/et"两句代码指定了其位置在 TextView 的下边、EditText 的右边。

"清空"按钮通过 android: layout_alignTop = "@+id/bt_ok"、android: layout_toRightOf="@+id/bt_ok"和 android:layout_marginLeft="25dp"三句代码指定了它的水平位置应该在"确定"按钮的右边,并距离后者 25dp 处,垂直位置应该与"确定"按钮顶部对齐,也就是水平对齐。

(2) Activity 组件 MainActivity 类代码如下。

```
package com.example.android_demo3_4

import android.app.Activity;
import android.os.Bundle;
import android.view.View;
import android.view.View.OnClickListener;
```

```java
import android.widget.Button;
import android.widget.EditText;
import android.widget.TextView;

/*主Activity为控制类,用于控制界面的显示*/
public class MainActivity extends Activity implements OnClickListener {
    private Button mButtonOK, mButtonCancel;
    private TextView mTextView;
    private EditText mEditText;

    /*Activity组件的生命周期方法,在组件创建时调用,一般用于初始化信息*/
    @Override
    protected void onCreate(Bundle savedInstanceState) {
        //调用父类方法,完成系统工作
        super.onCreate(savedInstanceState);
        //加载界面布局文件
        setContentView(R.layout.activity_main);
        //获取界面控件
        getViewItem();
    }

    /*获取界面控件*/
    private void getViewItem() {
        //根据控件id获取界面上的Button控件
        mButtonOK= (Button) findViewById(R.id.bt_ok);
        //根据控件id获取界面上的Button控件
        mButtonCancel= (Button) findViewById(R.id.bt_clear);
        //根据控件id获取界面上的TextView控件
        mTextView= (TextView) findViewById(R.id.tv);
        //根据控件id获取界面上的EditText控件
        mEditText= (EditText) findViewById(R.id.et);
        //给按钮设置单击事件
        mButtonOK.setOnClickListener(this);
        //给按钮设置单击事件
        mButtonCancel.setOnClickListener(this);
    }

    /*复写监听器对象的onClick方法,完成单击后的事件处理,参数为被单击的按钮*/
    @Override
    public void onClick(View v) {
        //根据按钮控件的id区分不同按钮的单击
        switch (v.getId()) {
        case R.id.bt_ok:
            //获取界面控件EditText的输入内容
```

```
            String _Text=mEditText.getText().toString();
            //将界面控件 TextView 的文本设置为输入内容
            mTextView.setText(_Text);
            break;
        case R.id.bt_clear:
            //清空界面控件 EditText 的文本输入内容
            mEditText.setText("");
            break;
        }
    }
}
```

5. 案例运行效果

案例运行效果如图 3-8 所示，图 3-8(a)显示的是没有在编辑框中输入内容的界面效果，图 3-8(b)显示的是在编辑框中输入内容后，并单击"确认"按钮的界面效果。

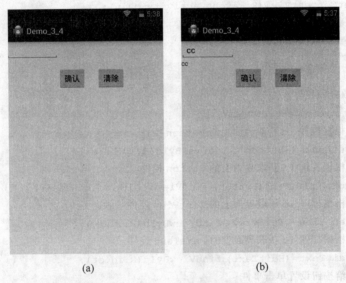

图 3-8　RelativeLayout 布局案例运行效果

3.2.3　表格布局

TableLayout 属于行和列形式的管理控件，适用于多行多列的布局格式，每行为一个 TableRow 对象，也可以是一个 View 对象。在 TableRow 中还可以继续添加其他的控件，每添加一个子控件就成为一列。TableLayout 不会生成边框，它的几个重要属性如表 3-5 所示。

表格布局的风格跟 HTML 中的表格比较接近，只是所采用的标签不同。<TableLayout>是顶级元素，说明采用的是表格布局，<TableRow>定义一个行，而具体控件则定义一个单元格的内容。

表 3-5 TableLayout 的重要属性表

XML 属性名	说明
Android:collapseColumns	设置指定的列为 collapse，如果一列被标示为 collapse，该列会被隐藏
Android:shrinkColumns	设置指定的列为 shrinkable，如果一列被标示为 shrinkable，列的宽度进行收缩，自适应父容器的大小
Android:stretchColumns	设置指定的列为 stretchable，如果一列被标示为 stretchable，该列会被拉伸，填充满表格的空白区域

下面通过分析一个 XML 布局文件来说明 TableLayout 的使用方法。有布局文件 activity_main.xml，其代码如下。

```xml
<?xml version="1.0" encoding="utf-8"?>
<!--最外面的布局文件为表格布局 -->
<TableLayout xmlns:android="http://schemas.android.com/apk/res/android"
    android:layout_width="fill_parent"
    android:layout_height="fill_parent"
    android:stretchColumns="0,1,2,3">
    <!--表格布局中第一行 TableRow -->
    <TableRow>
        <!--第一行的第一列控件 TextView -->
        <TextView
            android:gravity="center"
            android:padding="3dip"
            android:text="姓名" />
        <!--第一行的第二列控件 TextView -->
        <TextView
            android:gravity="center"
            android:padding="3dip"
            android:text="性别" />
        <!--第一行的第三列控件 TextView -->
        <TextView
            android:gravity="center"
            android:padding="3dip"
            android:text="年龄" />
        <!--第一行的第四列控件 TextView -->
        <TextView
            android:gravity="center"
            android:padding="3dip"
            android:text="电话" />
    </TableRow>
    <!--表格布局中第二行 TableRow -->
    <TableRow>
```

```xml
<!--第二行的第一列控件 TextView -->
<TextView
    android:gravity="center"
    android:padding="3dip"
    android:text="小明" />
<!--第二行的第二列控件 TextView -->
<TextView
    android:gravity="center"
    android:padding="3dip"
    android:text="男" />
<!--第二行的第三列控件 TextView -->
<TextView
    android:gravity="center"
    android:padding="3dip"
    android:text="20" />
<!--第二行的第四列控件 TextView -->
<TextView
    android:gravity="center"
    android:padding="3dip"
    android:text="180080000008" />
</TableRow>
<!--表格布局中第三行 TableRow -->
<TableRow>
    <!--第三行的第一列控件 TextView -->
    <TextView
        android:gravity="center"
        android:padding="3dip"
        android:text="小王" />
    <!--第三行的第二列控件 TextView -->
    <TextView
        android:gravity="center"
        android:padding="3dip"
        android:text="男" />
    <!--第三行的第三列控件 TextView -->
    <TextView
        android:gravity="center"
        android:padding="3dip"
        android:text="24" />
    <!--第三行的第四列控件 TextView -->
    <TextView
        android:gravity="center"
        android:padding="3dip"
        android:text="180000080000" />
```

```
        </TableRow>
    </TableLayout>
```

在上面的布局中实现了在界面上显示一个三行四列的表格,第一行是表头,显示"姓名"、"性别"、"年龄"和"电话",下面两行是具体信息。布局文件最外层容器为 TableLayout 布局,宽和高都是充满父容器,其中内嵌了三个 TableRow。在第一个 TableRow 中创建 4 个 TextView 为 4 列,都是居中对齐,内边距为 3dp,文本分别显示"姓名"、"性别"、"年龄"和"电话";在第二个 TableRow 中创建 4 个 TextView 为 4 列,都是居中对齐,内边距为 3dp,文本分别显示"小明"、"男"、"20"和"180080000008";在第三个 TableRow 中创建 4 个 TextView 为 4 列,都是居中对齐,内边距为 3dp,文本分别显示"小王"、"男"、"24"和"180000080000"。其中,android:stretchColumns="0,1,2,3"属性指定每行都由第 0、1、2、3 列占满空白空间。gravity 指定文字对齐方式,都设为居中对齐。padding 指定视图与视图内容间的空隙,单位为 dip,即独立设备像素,也可简写为 dp。上面布局文件的演示效果如图 3-9 所示。

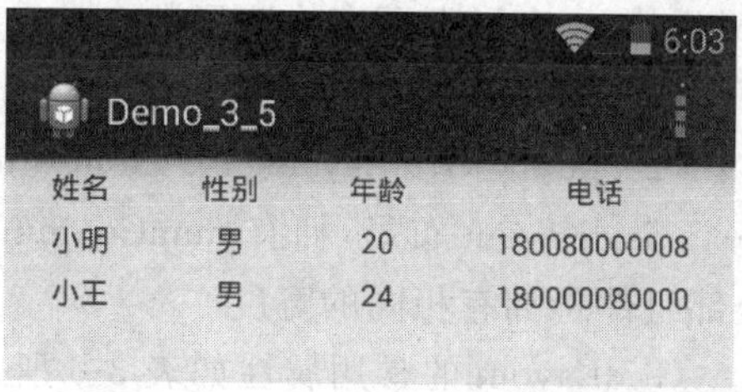

图 3-9 TableLayout 布局演示效果

3.2.4 帧布局

帧布局中的每一个组件都代表一个画面,默认以屏幕左上角作为(0,0)坐标,按组件定义的先后顺序依次逐屏显示,后面出现的会覆盖前面的画面。用该布局可以实现动画效果。一个使用帧布局的 XML 文件的代码如下。

```
<?xml version="1.0" encoding="utf-8"?>
<!--最外面的布局文件为帧布局 -->
<FrameLayout xmlns:android="http://schemas.android.com/apk/res/android"
    android:layout_width="fill_parent"
    android:layout_height="wrap_content">
    <!--帧布局中第一层控件 TextView -->
    <TextView
        android:layout_width="wrap_content"
        android:layout_height="wrap_content"
        android:text="A Text">
    </TextView>
    <!--帧布局中第一层控件 Button -->
    <Button
        android:layout_width="wrap_content"
        android:layout_height="wrap_content"
        android:text="A Button">
    </Button>
</FrameLayout>
```

上面布局文件的运行效果如图 3-10 所示。

从图 3-10 中可以看到,帧布局的各个控件会从屏幕的左上角,即(0,0)坐标开始,根据定义的先后顺序,逐个显示,于是 Button 和 TextView 就会如图 3-10 所示的那样,重叠显示。

图 3-10 FrameLayout 布局演示效果

3.2.5 网格布局

GridLayout 提供了一种新的布局方式,它可以将子视图放入到一个矩形网格中。GridLayout 有以下两个构造函数。

(1) public GridLayout():建立一个默认的 GridLayout 布局。

(2) public GridLayout(int numColumns, boolean makeColumnsEqualWidth):建立一个 GridLayout 布局,拥有 numColumns 列。如果 makeColumnsEqualWidth 为 true,则全部组件将拥有相同的宽度。

GridLayout 的常用属性如表 3-6 所示。

表 3-6 GridLayout 的常用属性

属性名称	属性含义
android:orientation	设置组件的排列方式,取值为 vertical(垂直)或 horizontal(水平)
android:layout_gravity	设置组件的对齐方式
android:rowCount	设置有多少行
android:columnCount	设置有多少列
android:layout_row	组件在第几行
android:layout_column	组件在第几列
android:layout_rowSpan	横跨几行
android:layout_columnSpan	横跨几列

GridLayout 中的元素一般不采用 layout_width 和 layout_height 来界定大小,而是采用 layout_gravity="fill_horizontal" 或 fill_vertical,并配合 GridLayout 的 android:orientation 属性来定义它里面的视图元素的大小。默认情况下,它里面的元素大小为 wrap_content。

GridLayout 中的 android:orientation 属性决定了其中的视图元素的摆放方式,如果为 vertical,则先摆第一列,然后第二列,以此类推;如果为 horizontal,则先摆第一行,然后第二行,以此类推。

一个使用 GridLayout 实现矩形网格的代码示例如下。

```
<?xml version="1.0" encoding="utf-8"?>
```

```xml
<GridLayout xmlns:android="http://schemas.android.com/apk/res/android"
    android:layout_width="wrap_content"
    android:layout_height="wrap_content"
    android:orientation="horizontal"
    android:rowCount="5"
    android:columnCount="4">

    <Button
        android:id="@+id/one"
        android:text="1" />
    <Button
        android:id="@+id/two"
        android:text="2"/>
    <Button
        android:id="@+id/three"
        android:text="3"/>
    <Button
        android:id="@+id/devide"
        android:text="/"/>
    <Button
        android:id="@+id/four"
        android:text="4"/>
    <Button
        android:id="@+id/five"
        android:text="5"/>
    <Button
        android:id="@+id/six"
        android:text="6"/>
    <Button
        android:id="@+id/multiply"
        android:text="×"/>
    <Button
        android:id="@+id/seven"
        android:text="7"/>
    <Button
        android:id="@+id/eight"
        android:text="8"/>
    <Button
        android:id="@+id/nine"
        android:text="9"/>
    <Button
        android:id="@+id/minus"
        android:text="-"/>
    <Button
```

```
        android:id="@+id/zero"
        android:layout_columnSpan="2"
        android:layout_gravity="fill"
        android:text="0"/>
<Button
        android:id="@+id/point"
        android:text="."/>
<Button
        android:id="@+id/plus"
        android:layout_rowSpan="2"
        android:layout_gravity="fill"
        android:text="+"/>
<Button
        android:id="@+id/equal"
        android:layout_columnSpan="3"
        android:layout_gravity="fill"
        android:text="="/>
</GridLayout>
```

上述布局代码的运行效果如图 3-11 所示。在上面的布局中顶层采用网格布局，android:layout_width="match_parent"和 android:layout_height="match_parent"表示网格布局宽度和高度将填充整个屏幕；android:orientation="horizontal"表示采用水平布局，采用了 5 行 4 列，按行一次排列，向 GridLayout 中放入 16 个按钮，其中按钮 0 跨越了两列、按钮＋跨越了两行、按钮＝跨越了三列。

图 3-11　GridLayout 布局演示效果

3.2.6　绝对布局

AbsoluteLayout，又可以叫作坐标布局，是直接按照控件的横纵坐标在界面中进行布局。因为考虑到不同屏幕分辨率的 Android 设备的适配问题，在 Google 的官方文档中，已经明确建议开发者不要使用绝对布局。由于所有的绝对布局都可以简单地通过其他的布局组合来完成,故不推荐读者使用这一布局方式。

绝对布局使用 android:layout_x 属性来确定 X 坐标,以左上角为顶点。使用 android:layout_y 属性确定 Y 坐标,以左上角为顶点。在绝对定位中,如果子元素不设置 layout_x 和 layout_y,那么它们的默认值是 0,也就是说它会像在 FrameLayout 中一样,这个元素会出现在屏幕的左上角。

习 题 3

1. 简述文本框(TextView)与编辑框(EditText)的功能,以及它们有什么区别。
2. 说明 Android 中常用的 5 种布局及其特点。
3. 下面的代码用来实现用户的登录验证功能,请在【1】、【2】、【3】号位置填入正确的代码。

```
package com.example.login;
import***
public class MainActivity extends Activity {
    private Button btnLogin;
    private EditText etUsername, etPassword;
    @Override
    protected void onCreate(Bundle savedInstanceState) {
        super.onCreate(savedInstanceState);
        setContentView(R.layout.activity_main);
        btnLogin= (Button) findViewById(R.id.button1);
        etUsername= (EditText) findViewById(R.id.editText1);
        etPassword= (EditText) findViewById(R.id.editText2);
        btnLogin.setOnClickListener(new OnClickListener() {
            @Override
            public void onClick(View v) {
                String username=    【1】    ;
                String password=    【2】    ;
                if (    【3】    ) {
                    return;
                }
                if (username.equals("admin") && password.equals("123456")) {
                    Toast.makeText(MainActivity.this,
                        "登录成功", Toast.LENGTH_LONG).show();
                }
            }
        });
    }
}
```

第 4 章

Android 活动简介

Activity 是 Android 的四大组件之一,是专门负责控制视图 View 与用户进行交互的活动类,也是在实际开发中使用频率最高的组件之一。Activity 组件从创建到销毁会经历不同的生命周期阶段,且在不同的生命周期阶段会对应回调不同的方法。我们虽然不能控制 Activity 的生命周期,但可以在对应的生命周期方法中完成特定的任务,例如在 Activity 创建的时候获取初始化数据、在 Activity 销毁的时候保存数据等。

Android 规定 Activity 组件必须在 AndroidManifest.xml 文件中通过＜activity＞和＜/activity＞标签进行注册,这样才可以通过解析 AndroidManifest.xml 中的＜activity＞标签找到对应的 Activity 进行启动。

Activity 与 Activity 所管理的界面之间是可以切换的,并且是可以在切换的过程中传递数据,甚至可以把数据从当前的 Activity 返回到第一个 Activity 中,这些工作需要一个类似信使的类来完成,这个信使就是 Intent(意图)。Intent 意图实际上可以在 Android 的不同组件间实现传递数据。Intent 可分为显示意图和隐式意图两类,其中隐式意图需要指定 Intent 的 Action 属性来指定意图的动作才能找到对应的组件。

Activity 的启动有 4 种模式,即 standard(默认模式)、singleTop、singleTask 和 singleInstance,可以在不同的场景下设置 AndroidManifext.xml 中的＜activity＞标签的 android:launchMode 属性来指定启动模式。

4.1 Activity 的创建与注册

4.1.1 Activity 的创建

在 Android 项目的 src 源代码目录中的包上右击,在弹出的快捷菜单中选择 New—＞Other,弹出的对话框如图 4-1 所示;在对话框中选择 Android—＞Android Activity,单击 Next 按钮;在如图 4-2 所示的对话框中选择一个 Activity 模板(本文选择默认),再单击 Next 按钮;最后在如图 4-3 所示的对话框中输入 Activity 类的名称、布局文件名称和标题,单击 Finish 按钮完成创建。

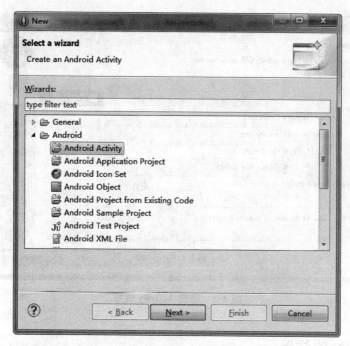

图 4-1　新建文件向导

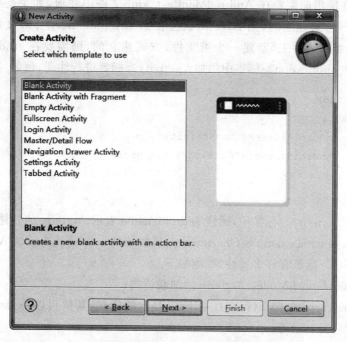

图 4-2　新建 Activity 模板选择

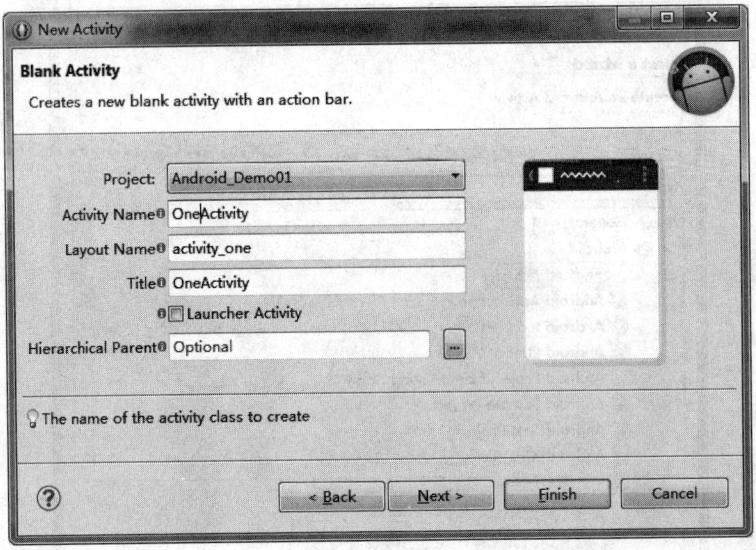

图 4-3 新建 Activity 设置

4.1.2 Activity 的注册

Activity 在使用前需要在 AndroidManifest.xml 文件中注册，如果是通过上面的新建 Activity 向导方式建立的，则 ADT 默认会在 AndroidManifest.xml 文件中注册该新建的 Activity；如果是手工（新建一个类文件）方式建立的，则需要在 AndroidManifest.xml 文件中的＜application＞标签中添加＜activity＞标签进行注册，例如：

```
<application>
    <activity
        android:name=".OneActivity"
        android:label="@string/title_activity_one">
    </activity>
</application>
```

在上面的＜activity＞标签中，属性 android:name 表示 Activity 的名称，采用 Java 风格的命名规范，如 com.example.OneActivity；属性 android:label 表示该 Activity 显示的标题。＜activity＞标签有许多属性，举例如下。

（1）android:launchMode：Activity 的加载模式；

（2）android:screenOrientation：Activity 显示的模式，其值可以是 landscae（横屏模式）、portrait（竖屏模式）等；

（3）android:icon：定义了代表 Activity 的一个图标，如果没有设置，就会使用给应用程序设置的图标来代替。

可以为＜activity＞标签指定多个过滤器，使用＜intent-filter＞（意图过滤器）指定。意图过滤器的目的是告诉其他组件如何启动这个 Activity。当使用 ADT 创建一个新工

程时,根 Activity 被自动创建,它已具有两个意图过滤器,例如:

```
<intent-filter>
    <action android:name="android.intent.action.MAIN" />
    <category android:name="android.intent.category.LAUNCHER"/>
</intent-filter>
```

<action>说明此 Activity 是应用程序的入口,<category>指出这个 Activity 需要在系统的应用列表中列出。如果想让你的 Activity 被其他程序调用,那么需要为它增加意图过滤器。这些意图过滤器包括<action>,<category>以及<data>标签,这些标签指明了你的 Activity 响应何种类型的 intent。

4.2 Activity 的生命周期

通过实现 Activity 的生命周期回调方法来管理 Activity,是创造既稳定又灵活的 Activity 的关键。Activity 的生命周期直接受到相关的其他 Activity 和它的任务以及所在栈的影响。一个 Activity 可生存在以下三种基本的状态中。

(1) Resumed:Activity 位于屏幕的最上层,并具有用户焦点,用户可以操作它,即运行状态。

(2) Paused:Activity B 位于最上层并获得输入焦点,Activity A 位于其下一层,但 Activity A 依然可见,此时 Activity A 就处于 Paused 状态。此状态的 Activity 依然是"活"的,因为它还是位于内存中,并且它被窗口管理器所管理。它只要获取到 CPU 时间片就可以接着运行,当然系统此时是不想让它运行的,所以不给它 CPU 时间片。此状态的 Activity 在 RAM 剩余极少时,可能被系统杀掉。

(3) Stoped:一个 Activity 如果被其他 Activity 完全遮盖,那么它就处于 Stoped 状态。此时它处于"后台"。此状态的 Activity 也是"活"的,它依然位于内存中,但是在窗口管理器中把它除名。然而,它不再被用户看到并且系统可以在其他组件需要内存时把它杀掉,也就是说它比 paused 状态的 Activity 更容易被杀死。

如果一个 Activity 处于 Paused 或 Stoped 状态,系统可以杀死它。杀死它的方法有比较温和的——请求 Activity 用 finish()自杀,或直接用暴力的方法——杀掉 Activity 所在的进程。不论怎样,Activity 被从内存中移除,当被杀或自杀的 Activity 重新启动时,它必须被从头创建。

Activity 包含 7 个生命周期方法,Activity 处于不同状态时会调用不同的方法。Activity 的生命周期方法回调过程如图 4-4 所示,其调用过程如下。

1. 启动 Activity

系统会先调用 onCreate()方法,然后调用 onStart()方法,最后调用 onResume(),Activity 进入运行状态。

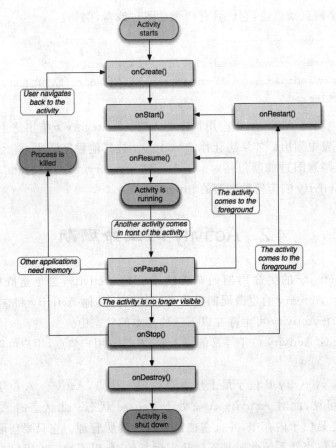

图 4-4 Activity 的生命周期方法回调过程

2. 当前 Activity 被其他 Activity 覆盖或被锁屏

系统会调用 onPause() 方法，暂停当前 Activity 的执行。

3. 当前 Activity 由被覆盖状态回到前台或解锁屏

系统会调用 onResume() 方法，再次进入运行状态。

4. 当前 Activity 转到新的 Activity 界面或按 Home 键回到主屏，自身退居后台

系统会先调用 onPause() 方法，然后调用 onStop() 方法，进入停滞状态。

5. 用户后退回到此 Activity

系统会先调用 onRestart() 方法，然后调用 onStart() 方法，最后调用 onResume() 方法，再次进入运行状态。

6. 当前 Activity 处于被覆盖状态或者后台不可见状态

即第 2 步和第 4 步，系统内存不足，杀死当前 Activity，而后用户退回当前 Activity，再次调用 onCreate() 方法、onStart() 方法和 onResume() 方法,进入运行状态。

7. 用户退出当前 Activity

系统先调用 onPause() 方法,然后调用 onStop() 方法,最后调用 onDestroy() 方法,结束当前 Activity。

表 4-1 详细地介绍了 Activity 的生命方法。

表 4-1　Activity 的生命方法

方　法	描　　述	完成后可杀掉	下一个
onCreate()	当 Activity 被创建时调用。一般的静态初始化工作可以在此方法中完成,例如创建界面,把数据绑定到列表等	否	onStart()
onRestart()	在停止后被调用,但不是停止后马上调用,而是在再次开始前调用,也就是在再次调用 onStart() 之前立即调用	否	onStart()
onStart()	当 Activity 变成可见后立即调用它。如果 Activity 成为最上层,则调用 onResume(),如果被完全遮盖,就调用 onStop()	否	onResume() 或 onStop()
onResume()	当 Activity 处于最上层时,立即调用此方法。此时 Activity 获得输入焦点	否	onPause()
onPause()	当另一个 Activity 要进入 Pause 状态时调用此方法。这个方法一般是用来提交那些发生改变的永久化的数据	是	onResume() 或 onStop()
onStop()	当 Activity 被完全遮盖时被调用。当 Activity 要销毁或被其他 Activity 完全遮盖时会发生	是	onRestart() 或 onDestroy()
onDestroy()	在 Activity 销毁之前被调用。这是 Activity 能收到的最后一个调用。调用的原因可能是这个 Activity 调用了 finish() 方法,也可能是系统为了更多的内存空间而把它所在的进程杀死了	是	nothing

4.3　Activity 的启动

4.3.1　直接启动 Activity

直接启动 Activity 可以使用 Activity 的 startActivity() 方法,但需要传递 Intent(意图)到方法的参数中,而且 Intent 必须指定了源 Activity 与目标 Activity。例如:

```
Intent intent=new Intent(this, SignInActivity.class);
startActivity(intent);
finish();
```

其中 SignInActivity 是要启动的 Activity,finish() 方法表示结束当前 Activity。

利用 Intent 直接启动 Activity 的另一种做法是通过 Intent 的 setClass() 方法来指定源 Activity 与目标 Activity,例如:

```
Intent intent=new Intent();
intent.setClass (LoginActivity.this, MainActivity.class);
startActivity(intent);
finish();
```

LoginActivity 是源 Activity，MainActivity 是要启动的 Activity，即从 LoginActivity 启动 MainActivity。

4.3.2　启动 Activity 并传递参数

Activity 之间要进行跳转和传递数据都需要 Intent（意图）介质类。在 Android 中提供了 Intent 机制来协助应用间的交互与通信，Intent 负责对应用中一次操作的动作、动作涉及的数据和附加数据进行描述，Android 则根据此 Intent 的描述，负责找到对应的组件，将 Intent 传递给调用的组件，并完成组件的调用。Intent 不仅可用于应用程序之间，也可用于应用程序内部的 Activity/Service 之间的交互。因此，可以将 Intent 理解为不同组件之间通信的"媒介"，它专门提供组件互相调用的相关信息。所以，如果要向新启动的 Activity 传递数据，就需要 Intent 的帮助，即将要传递的数据放入 Intent。

可以通过 Intent 类的 putExtra 方法将要传递的数据放入 Intent，该方法有多种重载。下面是常用的几种重载方式。

（1）Public Intent putExtra(String name,String value);
（2）Public Intent putExtra (String name,boolean value);
（3）Public Intent putExtra (String name,int value);
（4）Public Intent putExtra (String name,Serializable value);

其中，String name 为 key，第二个参数为 value，即 putExtra()按 key-value 形式放入数据。从上面多个构造方法可以看出，putExtra()方法可以保存各种类型的值。当通过 startActivity()方法启动新的 Activity 的时候，这些值也会一同传递到新启动的 Activity。在新的 Activity 的 onCreate()方法中，可以通过 getIntent.getxxxExtra()方法（getxxxExtra 中的 xxx 可以为 String、boolean、int 或 Serializable 类型）获取数据，举例如下。

从 LoginActivity 中启动 MainActivity，并传递数据：

```
Intent intent=new Intent(LoginActivity.this, MainActivity.class);
intent.putExtra("string_key", "string_value");
startActivity(intent);
```

在 MainActivity 接收数据：

```
Intent intent=getIntent();
String keyvalue=intent.getStringExtra("string_key");
```

也可以通过 bundle 对象来传递信息，bundle 维护了一个 HashMap<String,Object>对象，将数据存储在这个 HashMap 中来进行传递，举例如下。

从 LoginActivity 中启动 MainActivity,并传递数据:

```
Bundle bundle=new Bundle();
bundle.putString("string_key", "string_value");
intent.putExtra("key", bundle);
startActivity(intent);
```

在 MainActivity 接收数据:

```
Intent intent=getIntent();
Bundle bundle=intent.getBundleExtra("key");
String keyvalue=bundle.getString("string_key");
```

4.3.3 带返回值启动 Activity

所谓的带返回值启动 Activity,是指启动一个 Activity 后,在关闭这个新的 Activity 的时候可以将这个新 Activity 中必要的数据返回给启动它的那个 Activity。要想实现这样的功能,需要使用 Activity 的 startActivityForResult()方法来实现,但是必须重写 Activity 的 onActivityResult()方法才可以获得新 Activity 返回的数据。返回数据需要调用 setResult(int resultCode,Intent intent)方法将数据封装到 Intent 中,这样就可以在 onActivityResult(int resultCode,Intent data)方法中判断 resultCode 是否为新 Activity 返回的数据,可以通过 data 来获得返回的数据。举例如下。

在起始 Activity 中,发送数据的代码如下。

```
protected void onCreate(Bundle saveInstanceState){
    super.onCreate(saveInstanceState);
    setContentView(R.layout.loginactivity);
    Intent intent=new Intent();
    //设置起始 Activity 和目标 Activity,表示数据从这个 Activity 传到下个 Activity
    Intent intent=new Intent(LoginActivity.this, MainActivity.class);
    //绑定数据
    intent.putExtra("username",username);
    intent.putExtra("userpass",userpass);
    //打开目标 Activity
    startActivityForResult(intent,1);
}

//需要重写 onActivityResult 方法
protected void onActivityResult (int requestCode, int resultCode, Intent intent){
    super.onActivityResult(requestCode,resultCode,intent);
    //判断结果码是否与回传的结果码相同
    if(resultCode==1){
        //获取回传数据
```

```
        String name=intent.getStringExtra("name");
        String pass=intent.getStringExtra("pass");
        //对数据进行操作
        ...
    }
}
```

在目标 Activity 中,接收数据的代码如下。

```
protected void onCreate(Bundle saveInstanceState){
    super.onCreate(saveInstanceState);
    setContentView(R.layout.mainactivity);
    //获得意图
    Intent intent=getIntent();
    //读取数据
    String name=intent.getStringExtra("username");
    String pass=intent.getStringExtra("userpass");
    //编辑数据
    name=name+"is vaild";
    pass=pass+"is vaild";
    //数据发生改变,需要把改变后的值传递回原来的 Activity
    intent.putExtra("name",name);
    intent.putExtra("pass",pass);
    //setResult(int resultCode,Intent intent)方法
    setResult(1,intent);
    //销毁此 Activity,摧毁此 Activity 后将自动回到上一个 Activity
    finish();
}
```

4.4 Activity 的启动模式

启动模式简单地说就是 Activity 启动时的策略,在 AndroidManifest.xml 文件中的 <activity> 标签的 android:launchMode 属性可以设置 Activity 的启动模式。例如:

```
<activity
    android:name=".MainActivity"
    android:launchMode="standard" />
```

Activity 有 4 种启动模式,分别为 standard、singleTop、singleTask 和 singleInstance。可以根据实际的需求 Activity 设置对应的启动模式,从而可以避免创建大量重复的 Activity 等问题。

讲解启动模式之前,有必要先讲解一下"任务栈"的概念。每个应用程序都有一个任务栈,是用来存放 Activity 的,功能类似于函数调用的栈,先后顺序代表了 Activity 的出现顺

序。例如 Activity 的出现顺序为 Activity1 －＞ Activity2 －＞ Activity3，则任务栈如图 4-5 所示。下面分别讲解 Activity 的 4 种启动模式的作用。

1. standard

每次激活 Activity 时，都会创建 Activity 实例，并放入任务栈。例如任务栈中有 Activity1 和 Activity2，当 Activity2 每次被激活时，新的 Activity2 都会叠加在原来的 Activity2 上，如图 4-6 所示。

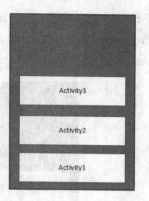

图 4-5　Activity 的任务栈

2. singleTop

如果某个 Activity 自己激活自己，即任务栈栈顶就是该 Activity，则不需要创建，其余情况都要创建 Activity 实例。例如任务栈中 Activity2 位于栈顶，Activity2 被激活后，则不需要重新创建，如图 4-7 所示。

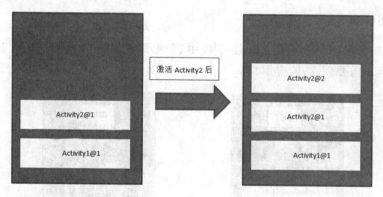

图 4-6　standard 模式示例

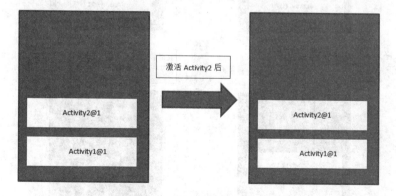

图 4-7　singleTop 模式示例

3. singleTask

如果要激活的那个 Activity 在任务栈中存在该实例，则不需要创建，只需要把此

Activity 放入栈顶,并把该 Activity 以上的 Activity 实例都弹出任务栈(销毁)。例如 Activity1 位于任务栈栈底,则 Activity1 被激活后,不需要再重新创建,但是需要将 Activity1 之上的所有 Activity,如图 4-8 所示的 Activity2 弹出栈。

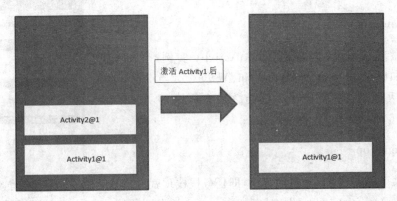

图 4-8 singleTask 模式示例

4. singleInstance

如图 4-9 所示,如果应用 2 的任务栈中创建了 Activity3 实例,应用 1 也要激活 Activity3,但不需要创建,两应用共享 Activity3 实例。

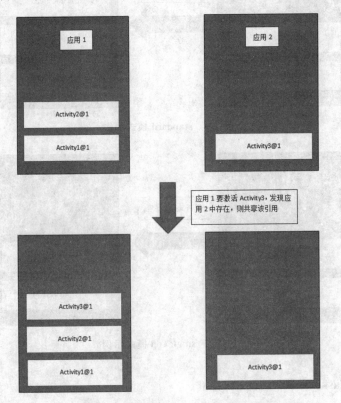

图 4-9 singleInstance 模式示例

4.5 案例 lifecycle

4.5.1 案例功能描述

案例将实现两个界面的跳转并互相传递参数,以及横纵屏切换的屏幕变化。

4.5.2 案例程序结构

案例中包括三个布局文件(lifecycle.xml、orientation_landscape.xml、orientation_portrait.xml),用于设计用户界面;两个 Activity 组件(LifeCycleActivity 类和 TargetActivity 类),用于实现用户界面交互功能。

4.5.3 案例的实现步骤和思路

1. 创建 Android 项目

具体操作略。

2. 创建两个 Activity

分别命名为 LifeCycleActivity 和 TargetActivity,在 AndroidManifest.xml 文件中注册,并设置 LifeCycleActivity 为启动界面。在注册 TargetActivity 时需注意,要在其属性中添加 android:configChanges="orientation|keyboardHidden|screenSize"。

3. 在 res 目录下的 layout 子目录中创建新的布局文件

(1) 创建 lifecycle.xml 文件,最外层容器为 LinearLayout 布局,宽和高都是充满父容器;再创建第二层容器为 LinearLayout 布局,里面包含一个 TextView 和一个 EditText 控件;并列再创建第二层容器为 LinearLayout 布局,里面也包含一个 TextView 和一个 EditText 控件;最后在第一层容器下创建一个 Button 控件。

(2) 在 res 目录下的 layout 子目录中创建新的布局文件 orientation_landscape.xml,最外层容器为 LinearLayout 布局,宽和高都是充满父容器,在 LinearLayout 布局内创建三个 TextView 控件和一个 Button 控件,LinearLayout 布局中最上面的 TextView 设置其文本显示为"横屏"。

(3) 在 res 目录下的 layout 子目录中创建新的布局文件 orientation_portrait.xml,将 orientation_landscape.xml 文件中的内容复制到此文件中,将 LinearLayout 布局中最上面的 TextView 的文本显示设置为"竖屏"。

4. 编写 LifeCycleActivity 类文件

(1) 复写生命周期方法 onCreate。使用 setContentView 方法加载布局文件 lifecycle.xml;获取界面控件 id,设置按钮监听,在按钮监听器的 onClick 方法中,取得用户在 EditText 中输入的用户名和密码放入 Intent,然后启动 TargetActivity 界面。

(2) 复写其他生命周期方法 onStart、onResume 等,在每个方法中使用 Log 输出日志。

5. 编写 TargetActivity 类文件

(1) 复写生命周期方法 onCreate。使用 setContentView 方法加载布局文件 orientation_portrait.xml;获取 LifeCycleActivity 传递的参数并显示;在按钮监听器的 onClick 方法中,返回数据给 LifeCycleActivity,结束自身 Activity。

(2) 复写其他生命周期方法 onStart、onResume 等,在每个方法中使用 Log 输出。

(3) 复写 onConfigurationChanged 方法,在方法中根据横屏和竖屏状态加载不同的布局。

4.5.4 案例参考代码

1. 布局文件 lifecycle.xml

```xml
<?xml version="1.0" encoding="utf-8"?>
<!--最外面的布局文件为线性布局 -->
<LinearLayout xmlns:android="http://schemas.android.com/apk/res/android"
    android:layout_width="match_parent"
    android:layout_height="match_parent"
    android:orientation="vertical">

    <LinearLayout
        android:layout_width="match_parent"
        android:layout_height="wrap_content"
        android:layout_margin="10dp">
        <TextView
            android:id="@+id/usertxt"
            android:layout_width="wrap_content"
            android:layout_height="wrap_content"
            android:layout_weight="1"
            android:text="用户名" />
        <EditText
            android:id="@+id/username"
            android:layout_width="wrap_content"
            android:layout_height="wrap_content"
            android:layout_weight="1"
            android:ems="10">
            <requestFocus />
        </EditText>
    </LinearLayout>
```

```xml
<LinearLayout
    android:layout_width="match_parent"
    android:layout_height="wrap_content"
    android:layout_margin="10dp">
    <TextView
        android:id="@+id/passtxt"
        android:layout_width="wrap_content"
        android:layout_height="wrap_content"
        android:layout_weight="1"
        android:text="密    码" />
    <EditText
        android:id="@+id/passwd"
        android:layout_width="wrap_content"
        android:layout_height="wrap_content"
        android:layout_weight="1"
        android:ems="10">
        <requestFocus />
    </EditText>
</LinearLayout>

<Button
    android:id="@+id/loginbtn"
    android:layout_width="match_parent"
    android:layout_height="wrap_content"
    android:layout_margin="10dp"
    android:text="登录" />

</LinearLayout>
```

2. 布局文件 orientation_landscape.xml

```xml
<?xml version="1.0" encoding="utf-8"?>
<!--最外面的布局文件为线性布局 -->
<LinearLayout xmlns:android="http://schemas.android.com/apk/res/android"
    android:layout_width="match_parent"
    android:layout_height="match_parent"
    android:layout_margin="10dp"
    android:orientation="vertical">

    <TextView
        android:id="@+id/ptxt"
        android:layout_width="wrap_content"
        android:layout_height="wrap_content"
        android:layout_gravity="top|center_horizontal"
```

```xml
            android:text="横屏"
            android:textSize="20sp" />

        <TextView
            android:id="@+id/loginusertxt"
            android:layout_width="wrap_content"
            android:layout_height="wrap_content"
            android:layout_margin="10dp"
            android:text=""
            android:textSize="15sp" />

        <TextView
            android:id="@+id/loginpasstxt"
            android:layout_width="wrap_content"
            android:layout_height="wrap_content"
            android:layout_margin="10dp"
            android:text=""
            android:textSize="15sp" />

    <Button
        android:id="@+id/validbtn"
        android:layout_width="match_parent"
        android:layout_height="wrap_content"
        android:layout_margin="10dp"
        android:text="验证" />

</LinearLayout>
```

3. 布局文件 orientation_portrait.xml

```xml
<?xml version="1.0" encoding="utf-8"?>
<!--最外面的布局文件为线性布局 -->
<LinearLayout xmlns:android="http://schemas.android.com/apk/res/android"
    android:layout_width="match_parent"
    android:layout_height="match_parent"
    android:layout_margin="10dp"
    android:orientation="vertical">

        <TextView
            android:id="@+id/ptxt"
            android:layout_width="wrap_content"
            android:layout_height="wrap_content"
            android:layout_gravity="top|center_horizontal"
            android:text="竖屏"
```

```xml
        android:textSize="20sp" />

    <TextView
        android:id="@+id/loginusertxt"
        android:layout_width="wrap_content"
        android:layout_height="wrap_content"
        android:layout_margin="10dp"
        android:text=""
        android:textSize="15sp" />

    <TextView
        android:id="@+id/loginpasstxt"
        android:layout_width="wrap_content"
        android:layout_height="wrap_content"
        android:layout_margin="10dp"
        android:text=""
        android:textSize="15sp" />

<Button
    android:id="@+id/validbtn"
    android:layout_width="match_parent"
    android:layout_height="wrap_content"
    android:layout_margin="10dp"
    android:text="验证" />

</LinearLayout>
```

4. Activity 组件 LifeCycleActivity 类

```java
package com.example.android_demo4_1;

import android.app.Activity;
import android.content.Context;
import android.content.Intent;
import android.os.Bundle;
import android.util.Log;
import android.view.View;
import android.widget.Button;
import android.widget.EditText;

/**
 * 主 Activity 为控制类,用于控制界面的显示
 */
public class LifeCycleActivity extends Activity {
```

```java
private static final String TAG="LifeCycleActivity";
private Context context=this;
private int param=1;
private EditText username;           //接收用户输入的用户名
private EditText passwd;             //接收用户输入的密码
//Activity 创建时被调用
@Override
public void onCreate(Bundle savedInstanceState) {
    super.onCreate(savedInstanceState);
    Log.i(TAG, "onCreate called.");
    setContentView(R.layout.lifecycle);
    username=(EditText) findViewById(R.id.username);
    passwd=(EditText) findViewById(R.id.passwd);
    Button btn= (Button) findViewById(R.id.loginbtn);
    btn.setOnClickListener(new View.OnClickListener() {
        @Override
        public void onClick(View v) {
            Intent intent=new Intent(context, TargetActivity.class);
            //绑定数据
            intent.putExtra("username",username.getText().toString());
            intent.putExtra("passwd",passwd.getText().toString());
            //打开目标 Activity
            startActivityForResult(intent,1);
        }
    });
}

//Activity 创建或者从后台重新回到前台时被调用
@Override
protected void onStart() {
    super.onStart();
    Log.i(TAG, "onStart called.");
}

//Activity 从后台重新回到前台时被调用
@Override
protected void onRestart() {
    super.onRestart();
    Log.i(TAG, "onRestart called.");
}

//Activity 创建或者从被覆盖、后台重新回到前台时被调用
@Override
protected void onResume() {
```

```java
        super.onResume();
        Log.i(TAG, "onResume called.");
}


//Activity窗口获得或失去焦点时被调用,在onResume之后或onPause之后
@Override
public void onWindowFocusChanged(boolean hasFocus) {
        super.onWindowFocusChanged(hasFocus);
        Log.i(TAG, "onWindowFocusChanged called.");
}


//Activity被覆盖到下面或者锁屏时被调用
@Override
protected void onPause() {
        super.onPause();
        Log.i(TAG, "onPause called.");
        //有可能在执行完onPause或onStop后,系统资源紧张将Activity杀死,所以
        //有必要在此保存持久数据
}


//退出当前Activity或者跳转到新Activity时被调用
@Override
protected void onStop() {
        super.onStop();
        Log.i(TAG, "onStop called.");
}


//退出当前Activity时被调用,调用之后Activity就结束了
@Override
protected void onDestroy() {
        super.onDestroy();
        Log.i(TAG, "onDestory called.");
}


@Override
protected void onActivityResult(int requestCode, int resultCode, Intent data) {
    //TODO Auto-generated method stub
    super.onActivityResult(requestCode, resultCode, data);
    //判断结果码是否与回传的结果码相同
    if(resultCode==1){
        //获取回传数据
        String name=data.getStringExtra("name");
        String pass=data.getStringExtra("pass");
        //对回传数据进行显示
```

```java
            username.setText(name);
            passwd.setText(pass);
        }
    }

    /**
     * Activity 被系统杀死时被调用
     * 例如:屏幕方向改变时,Activity 被销毁再重建;当前 Activity 处于后台,系统资源紧
     *   张将其杀死
     * 另外,当跳转到其他 Activity 或者按 Home 键回到主屏时该方法也会被调用,系统是为
     *   了保存当前 View 组件的状态在 onPause 之前被调用
     */
    @Override
    protected void onSaveInstanceState(Bundle outState) {
        outState.putInt("param", param);
        Log.i(TAG, "onSaveInstanceState called. put param: "+param);
        super.onSaveInstanceState(outState);
    }

    /**
     * Activity 被系统杀死后再重建时被调用
     * 例如:屏幕方向改变时,Activity 被销毁再重建;当前 Activity 处于后台,系统资源紧
     *   张将其杀死,用户又启动该 Activity
     * 这两种情况下 onRestoreInstanceState 都会在 onStart 之后被调用
     */
    @Override
    protected void onRestoreInstanceState(Bundle savedInstanceState) {
        param=savedInstanceState.getInt("param");
        Log.i(TAG, "onRestoreInstanceState called. get param: "+param);
        super.onRestoreInstanceState(savedInstanceState);
    }
}
```

5. Activity 组件 TargetActivity 类

```java
package com.example.android_demo4_1;

import android.app.Activity;
import android.content.Intent;
import android.content.res.Configuration;
import android.os.Bundle;
import android.util.Log;
import android.view.View;
import android.widget.Button;
```

```java
import android.widget.TextView;
/**
 * 主Activity为控制类,用于控制界面的显示
 */
public class TargetActivity extends Activity {
    private static final String TAG="OrientationActivity";
    private int param=1;
    private String name="";
    private String pass="";
    TextView username=null;
    TextView passwd=null;
    Button btn=null;
    Intent intent=null;
    @Override
    protected void onCreate(Bundle savedInstanceState) {
        super.onCreate(savedInstanceState);
        setContentView(R.layout.orientation_portrait);
        Log.i(TAG, "onCreate called.");
        //获得意图
        intent=getIntent();
        //读取数据
        name=intent.getStringExtra("username");
        pass=intent.getStringExtra("passwd");
        init();
    }

    private void init() {
        Log.i(TAG, "Init called.");
        username=(TextView) findViewById(R.id.loginusertxt);
        passwd=(TextView) findViewById(R.id.loginpasstxt);
        btn=(Button) findViewById(R.id.validbtn);
        //显示数据
        username.setText("Login Username : "+name);
        passwd.setText("Login Password : "+pass);
        //设置按钮监听事件
        btn.setOnClickListener(new View.OnClickListener() {
            @Override
            public void onClick(View v) {
                //编辑数据
                name=name+" is valid";
                pass=pass+" is valid";
                //数据发生改变,需要把改变后的值传递回原来的Activity
                intent.putExtra("name",name);
                intent.putExtra("pass",pass);
```

```java
                    //setResult(int resultCode,Intent intent)方法
                    setResult(1,intent);
                    //销毁此 Activity,销毁此 Activity 后将自动回到上一个 Activity
                    finish();
                }
            });
    }
    @Override
    protected void onStart() {
        super.onStart();
        Log.i(TAG, "onStart called.");
    }
    @Override
    protected void onRestart() {
        super.onRestart();
        Log.i(TAG, "onRestart called.");
    }
    @Override
    protected void onResume() {
        super.onResume();
        Log.i(TAG, "onResume called.");
    }
    @Override
    protected void onPause() {
        super.onPause();
        Log.i(TAG, "onPause called.");
    }
    @Override
    protected void onStop() {
        super.onStop();
        Log.i(TAG, "onStop called.");
    }
    @Override
    protected void onDestroy() {
        super.onDestroy();
        Log.i(TAG, "onDestory called.");
    }

    @Override
    protected void onSaveInstanceState(Bundle outState) {
        outState.putInt("param", param);
        Log.i(TAG, "onSaveInstanceState called. put param: "+param);
        super.onSaveInstanceState(outState);
    }
```

```
@Override
protected void onRestoreInstanceState(Bundle savedInstanceState) {
    param=savedInstanceState.getInt("param");
    Log.i(TAG, "onRestoreInstanceState called. get param: "+param);
    super.onRestoreInstanceState(savedInstanceState);
}

//当指定了 android:configChanges="orientation"后
//方向改变时 onConfigurationChanged 被调用
@Override
public void onConfigurationChanged(Configuration newConfig) {
    super.onConfigurationChanged(newConfig);
    Log.i(TAG, "onConfigurationChanged called.");
    switch (newConfig.orientation) {
    case Configuration.ORIENTATION_PORTRAIT:
        setContentView(R.layout.orientation_portrait);
        init();
        break;
    case Configuration.ORIENTATION_LANDSCAPE:
        setContentView(R.layout.orientation_landscape);
        init();
        break;
    }
}
}
```

4.5.5 案例运行效果

案例启动后 LifeCycleActivity 的界面如图 4-10 所示，在用户输入"用户名"和"密码"并单击"登录"按钮之后，就会跳转到如图 4-11 所示的 TargetActivity 界面。可以看到，用户在 LifeCycleActivity 界面中输入的内容已经传递到 TargetActivity 中，并显示出来。

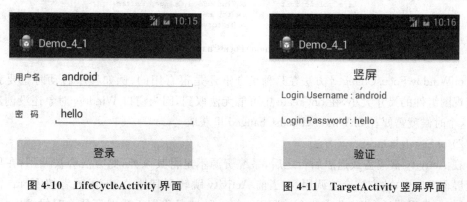

图 4-10　LifeCycleActivity 界面　　　　图 4-11　TargetActivity 竖屏界面

当用户在 TargetActivity 中旋转屏幕时,程序界面会自动更改布局,如图 4-12 所示,在用户单击"验证"按钮后,TargetActivity 结束返回到 LifeCycleActivity 界面,如图 4-13 所示。在 TargetActivity 结束前会将已经更改的数据放入到 Intent,在 LifeCycleActivity 的 onActivityResult 方法中取得返回的数据并显示到 EditText 中。

图 4-12 TargetActivity 横屏界面

图 4-13 返回到 LifeCycleActivity 界面

4.5.6 案例程序分析

1. LifeCycleActivity 类

在 LifeCycleActivity 类中,除了几个常见的方法外,我们还添加了 onWindowFocusChanged、onSaveInstanceState 和 onRestoreInstanceState 方法,现说明如下。

1) onWindowFocusChanged 方法

Activity 窗口获得或失去焦点时被调用,例如创建时首次呈现在用户面前;当前 Activity 被其他 Activity 覆盖;当前 Activity 转到其他 Activity 或按 Home 键回到主屏,自身退居后台;用户退出当前 Activity。以上几种情况都会调用 onWindowFocusChanged 方法,并且当 Activity 被创建时在 onResume 之后被调用,当 Activity 被覆盖或者退居后台或者当前 Activity 退出时,在 onPause 之后被调用,如图 4-14 所示。

```
LifeCycleActivity    onResume called.
LifeCycleActivity    onWindowFocusChanged called.
LifeCycleActivity    onPause called.
LifeCycleActivity    onWindowFocusChanged called.
LifeCycleActivity    onStop called.
LifeCycleActivity    onDestory called.
```

图 4-14 onWindowFocusChanged 方法调用

onWindowFocusChanged 方法在某种场合下还是很有用的,例如程序启动时想要获取特定视图组件的尺寸大小,在 onCreate 中可能无法取到,因为窗口 Window 对象还没创建完成,这个时候就需要在 onWindowFocusChanged 里获取。

2) onSaveInstanceState 方法

在 Activity 被覆盖或退居后台之后,系统资源不足将其杀死,此方法会被调用;在用户改变屏幕方向时,此方法会被调用;在当前 Activity 跳转到其他 Activity 或者按 Home 键回到主屏,自身退居后台时,此方法会被调用。第一种情况我们无法保证什么时候发生,系统根据资源紧张程度去调度;第二种是屏幕翻转方向时,系统先销毁当前的 Activity,然后再重建一个新的,调用此方法时,我们可以保存一些临时数据;第三种情况系统调用此方法是为了保存当前窗口各个 View 组件的状态。onSaveInstanceState 的调用顺序是在 onPause 之前。

3) onRestoreInstanceState 方法

在 Activity 被覆盖或退居后台之后,系统资源不足将其杀死,然后用户又回到了此 Activity,此方法会被调用;在用户改变屏幕方向时,重建的过程中,此方法会被调用。我们可以重写此方法,以便恢复一些临时数据。onRestoreInstanceState 的调用顺序是在 onStart 之后。

以上着重介绍了三个相对陌生的方法之后,下面就来运行这个 Activity,看看它的生命周期过程。

(1) 启动 Activity,LogCat 日志显示如图 4-15 所示。

tag	Message
LifeCycleActivity	onCreate called.
LifeCycleActivity	onStart called.
LifeCycleActivity	onResume called.

图 4-15　Activity 启动时的 LogCat 日志

在系统调用了 onCreate 和 onStart 之后,调用了 onResume,自此,Activity 进入了运行状态。

(2) 跳转到其他 Activity,或按 Home 键回到主屏,LogCat 日志显示如图 4-16 所示。

tag	Message
LifeCycleActivity	onSaveInstanceState called. put param: 1
LifeCycleActivity	onPause called.
LifeCycleActivity	onStop called.

图 4-16　Activity 退到后台时的 LogCat 日志

可以看到,此时 onSaveInstanceState 方法在 onPause 之前被调用了,并且注意,退居后台时,onPause 后 onStop 被调用。

(3) 从后台回到前台,LogCat 日志显示如图 4-17 所示。

tag	Message
LifeCycleActivity	onRestart called.
LifeCycleActivity	onStart called.
LifeCycleActivity	onResume called.

图 4-17　Activity 返回到前台时的 LogCat 日志

当从后台回到前台时,系统先调用 onRestart 方法,然后调用 onStart 方法,最后调用 onResume 方法,Activity 又进入了运行状态。

(4) 退出,LogCat 日志显示如图 4-18 所示。最后 onDestory 方法被调用,标志着 LifeCycleActivity 的终结。

tag	Message
LifeCycleActivity	onPause called.
LifeCycleActivity	onStop called.
LifeCycleActivity	onDestory called.

图 4-18　Activity 退出时的 LogCat 日志

2. TargetActivity 类

在 LifeCycleActivity 的运行过程中，并没有 onRestoreInstanceState 方法出现，原因是，onRestoreInstanceState 只有在杀死不在前台的 Activity 之后，用户回到此 Activity，或者用户改变屏幕方向的这两个重建过程中被调用。

为了演示 onRestoreInstanceState 方法的调用过程，先将 AndroidMainfest.xml 中的对 TargetActivity 的属性配置 android:configChanges="orientation|keyboardHidden|screenSize"注释掉，然后在 LifeCycleActivity 界面中，当我们旋转屏幕时，发现系统会先将当前 Activity 销毁，然后重建一个新的 Activity，LogCat 日志如图 4-19 所示。

tag	Message
OrientationActivity	onSaveInstanceState called. put param: 1
OrientationActivity	onPause called.
OrientationActivity	onStop called.
OrientationActivity	onDestory called.
OrientationActivity	onCreate called.
OrientationActivity	onStart called.
OrientationActivity	onRestoreInstanceState called. get param: 1
OrientationActivity	onResume called.

图 4-19　onRestoreInstanceState 方法调用过程

系统先是调用 onSaveInstanceState 方法，保存了一个临时参数到 Bundle 对象里面，然后当 Activity 重建之后我们又成功地取出了这个参数。

为了避免这样销毁重建的过程，需要在 AndroidMainfest.xml 中对 Activity 对应的属性进行配置，即添加属性 android:configChanges="orientation|keyboardHidden|screenSize"(将上面 AndroidMainfest.xml 中的注释去掉)，然后再测试，屏幕旋转了 4次，测试之后的 LogCat 日志如图 4-20 所示。

tag	Message
OrientationActivity	onConfigurationChanged called.
OrientationActivity	onConfigurationChanged called.
OrientationActivity	onConfigurationChanged called.
OrientationActivity	onConfigurationChanged called.

图 4-20　onConfigurationChanged 方法调用过程

可以看到，每次旋转方向时，只有 onConfigurationChanged 方法被调用，没有了销毁重建的过程。

需要注意，如果＜activity＞配置了 android:screenOrientation 属性，则会使 android:configChanges="orientation|keyboardHidden|screenSize"属性配置失效。

习 题 4

1. 请描述 Activity 的生命周期所调用的方法。
2. 横竖屏切换时，Activity 的生命周期是什么？
3. 什么是 Intent？它有什么作用？
4. Activity 有哪些启动模式？

第 5 章

Android 高级 UI 编程

在第 2 章我们学习了 Android 界面最基本的 UI 控件,也是使用频率最高的 UI 控件,在开发一个 Android 应用中会大量地使用这些最基本的控件,但是要实现一些类似于 QQ 好友列表、京东商品列表这样的界面时,基本的 UI 控件可能就满足不了我们的需求了。本章将要介绍一些 Android 的高级 UI 控件,可以通过这些控件,制作出更为复杂、功能更强大的 UI,增强 Android 应用的用户体验。

5.1 Adapter 简介

在 Android 的高级控件中有一类控件数据 AdapterView,这些控件有一个共同特点就是都需要 Adapter 适配器来为控件生成条目 View,并为每个条目 View 绑定数据。例如常用的 ListView、GridView 和 Spinner 等都属于 AdapterView。在我们所用到的 Adapter 中有简单的,也有复杂的,我们就从简单的 Adapter 开始来学习这些高级 UI 控件的一般用法。

Adapter 是连接后端数据和前端显示的适配器接口,是数据和 UI(View)之间一个重要的纽带。常见的 View(ListView、GridView)都需要用到 Adapter。图 5-1 直观地表达了 Data Source、Adapter、View 三者的关系,Adapter 将 Data Source 与 UI 显示(ListView)联系在一起。

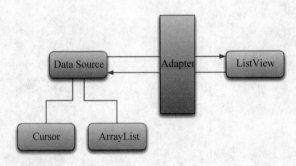

图 5-1 Data Source、Adapter、View 三者的关系

Adapter 其实只是一个接口,有一些常用的不同的实现类。Android 中常用的

Adapter 如下。

（1）BaseAdapter：基础适配器，实现了 ListAdapter 和 SpinnerAdapter 的接口。GridView 的适配器实现了 ListAdapter，所以，BaseAdapter 对于 Spinner、ListView 和 GridView 是通用的。BaseAdapter 是一个抽象类，继承它的类要实现很多方法。

（2）ArrayAdapter：数组适配器，BaseAdapter 的子类，从 BaseAdapter 派生出来，它有 BaseAdapter 的所有功能。但是，ArrayAdapter 可以直接使用泛型结构。ArryAdapter 比较简单，只能显示一行文本。

（3）SimpleAdapter：简单适配器，可以将静态数据映射到 XML 文件定义好的布局中。利用 SimpleAdapter 可以显示比较复杂的列表。例如，每行显示一些图片、文本等一些复杂特殊的效果，但它只是单纯地显示，若要在后期对列表进行修改，则不可以。SimpleAdapter 具有最后的扩充性。

（4）SimpleCursorAdapter：是一个专门用于将数据库表中的数据显示在 UI 组件中的适配器。在 Android 中一些普通的 Adapter 对象也可以将数据库中的数据显示在界面上，但用普通的 Adapter 工作量要大很多。SimpleCursorAdapter 在使用中主要是将 Cursor 的字段与 UI 的 ID 对应起来。

5.1.1　ArrayAdapter 简介

ArrayAdapter 一般用于显示一行文本信息。它接收一个数组或者 List 作为参数来构建。ArrayAdapter 的构造函数如下：

```
public ArrayAdapter (Context context, int resource, List<T>objects);
```

其参数含义如下。

（1）Context context：运行着的视图的上下文；

（2）int resource：一个包含有一个 TextView 布局的资源唯一标识，这个布局文件可以自己编写，也可以使用系统中的，如 android. R. layout. simple_expandable_list_item_1（显示一行文本的布局）；

（3）List<T> objects：一个 List 集合，也就是数据源。

ArrayAdapter 的使用方法是：先根据一个布局文件和一个 List 集合构造一个 ArrayAdapter 实例，然后通过控件的 setAdapter（）方法绑定 ArrayAdapter，例如 ListView 控件。

下面通过简单的例子进行说明。

5.1.2　案例 ArrayAdapter

1. 案例功能描述

ArrayAdapter 是最简单的 ListView 的练习案例，案例将实现在界面上以列表形式展示几条数据。

2. 案例程序结构

案例中包括一个 Activity 组件（MainActivity 类），用于实现用户界面交互功能。

3. 案例的实现步骤和思路

（1）创建 Android 项目。
（2）编写 MainActivity 文件，复写生命周期方法 onCreate。
① 先调用父类的 onCreate 方法；
② 创建 ListView 对象，并赋值给成员变量 lv；
③ 编写方法 getData，在方法中创建一个 List 集合，存放几条数据到集合中然后返回；
④ 给 ListView 设置适配器，使用 ArrayAdapter，并调用 getData 方法把数据源传入；
⑤ 最后调用 setContentView 方法将 ListView 控件显示在界面上。

4. 案例参考代码

Activity 组件 MainActivity 类的代码如下。

```java
package com.example.android_demo5_1

import java.util.ArrayList;
import java.util.List;

import android.app.Activity;
import android.os.Bundle;
import android.widget.ArrayAdapter;
import android.widget.ListView;
/**
 * 主 Activity 为控制类,用于控制界面的显示
 */
public class MainActivity extends Activity {
    //ListView 控件
    private ListView lv;
    @Override
    protected void onCreate(Bundle savedInstanceState) {
        //调用父类方法,完成系统工作
        super.onCreate(savedInstanceState);
        //创建 ListView 对象
        lv=new ListView(this);
        //给 ListView 设置适配器
        lv.setAdapter(new ArrayAdapter<String>(this,
            android.R.layout.simple_expandable_list_item_1,getData()));
```

```
        //将 ListView 控件显示在界面上
        setContentView(lv);
    }
    /**
     * 创建数据源数据,并返回
     * @return
     */
    private List<String>getData(){

        List<String>data=new ArrayList<String>();
        data.add("测试数据 1");
        data.add("测试数据 2");
        data.add("测试数据 3");
        data.add("测试数据 4");
        return data;
    }
}
```

5. 案例运行效果

ArrayAdapter 案例的运行效果如图 5-2 所示。

5.1.3 SimpleAdapter 简介

SimpleAdapter 是一个简单的适配器,可以将静态数据映射到 XML 文件中定义好的视图。SimpleAdapter 可以指定数据支持的列表,如 ArrayList 组成的 Map,在 ArrayList 中的每个条目对应 List 中的一行,Maps 包含每行数据。在构造 SimpleAdapter 时可以指定一个定义了被用于显示行的视图 XML 文件,通过关键字映射到指定的视图。

图 5-2　ArrayAdapter 案例运行效果

SimpleAdapter 的构造函数如下:

```
public SimpleAdapter (Context context, List<? extends Map<String, ?>>data, int resource, String[] from, int[] to);
```

其构造函数参数说明如下。

(1) context:关联 SimpleAdapter 运行着的视图的上下文。

(2) data:一个 Map 列表。在列表中的每个条目对应列表中的一行,应该包含所有在 from 中指定的条目。

(3) resource:一个定义列表项目的视图布局的资源唯一标识。布局文件至少应包含那些在 to 中定义了的名称。

(4) from:一个将被添加到 Map 上关联每一个项目的列名称的列表。

(5) to:应该在参数 from 显示列的视图,应该全是 TextView。在列表中最初的

N 视图是从参数 from 中最初的 N 列获取的值。下面以案例来说明 SimpleAdapter 的用法。

5.1.4 案例 SimpleAdapter

1. 案例功能描述

案例将实现在界面上以列表形式展示几条数据，数据稍微复杂些，左边是图片，右边是文本信息。

2. 案例程序结构

案例中包括两个布局文件（activity_main.xml、listitem.xml），用于设计用户界面和显示 ListView 每个条目的布局；一个 Activity 组件（MainActivity 类），用于实现用户界面交互功能。

3. 案例的实现步骤和思路

（1）创建 Android 项目。

（2）在 res 目录下的 layout 子目录中创建新的布局文件 activity_main.xml，最外层容器为 RelativeLayout 布局，宽和高都是充满父容器；创建内嵌 ListView，宽和高为充满父容器，id 为 lv。

（3）在 res 目录下的 layout 子目录中创建新的布局文件 listitem.xml，是 ListView 每个条目的布局，最外层容器为 RelativeLayout 布局，宽是充满父容器，高是包裹内容，在其内创建一个 ImageView，宽和高为充满包裹内容，id 为 iv，再创建一个 TextView，宽和高为充满包裹内容，id 为 tv。

（4）编写 MainActivity 文件，复写生命周期方法 onCreate。

① 先调用父类的 onCreate 方法；

② 使用 setContentView 方法加载布局文件 activity_main.xml；

③ 根据界面控件 id 使用 findViewById 方法，获取界面控件 ListView 并赋值给局部变量 lv；

④ 创建数据源集合 List＜Map＜String,Object＞＞data＝new ArrayList＜Map＜String,Object＞＞()；

⑤ 创建 5 条 Map 数据放入数据源集合中；

⑥ 给界面控件 lv 设置适配器。

4. 案例参考代码

（1）布局文件 activity_main.xml 的代码如下。

```
<?xml version="1.0" encoding="utf-8"?>
<!--主界面布局,最外面的布局文件为相对布局 -->
<RelativeLayout xmlns:android="http://schemas.android.com/apk/res/android"
```

```xml
    xmlns:tools="http://schemas.android.com/tools"
    android:layout_width="match_parent"
    android:layout_height="match_parent"
    android:paddingBottom="@dimen/activity_vertical_margin"
    android:paddingLeft="@dimen/activity_horizontal_margin"
    android:paddingRight="@dimen/activity_horizontal_margin"
    android:paddingTop="@dimen/activity_vertical_margin"
    tools:context=".MainActivity">
    <!--布局中第一个控件 ListView -->
    <ListView
        android:id="@+id/lv"
        android:layout_width="match_parent"
        android:layout_height="match_parent"/>

</RelativeLayout>
```

(2) ListView 条目布局的布局文件 listitem.xml 的代码如下。

```xml
<?xml version="1.0" encoding="utf-8"?>
<!--ListView 每个条目的布局,最外面的布局文件为线程布局 -->
<LinearLayout xmlns:android="http://schemas.android.com/apk/res/android"
    android:layout_width="match_parent"
    android:layout_height="wrap_content"
    android:layout_gravity="center_vertical"
    android:orientation="horizontal">
    <!--布局中第一个控件 ImageView -->
    <ImageView
        android:id="@+id/iv"
        android:layout_width="wrap_content"
        android:layout_height="wrap_content" />
    <!--布局中第二个控件 TextView -->
    <TextView
        android:id="@+id/tv"
        android:layout_width="wrap_content"
        android:layout_height="wrap_content" />
</LinearLayout>
```

(3) Activity 组件 MainActivity 类代码如下。

```java
package com.example.android_demo5_2

import java.util.ArrayList;
import java.util.HashMap;
import java.util.List;
import java.util.Map;
```

```java
import android.app.Activity;
import android.os.Bundle;
import android.widget.ListView;
import android.widget.SimpleAdapter;
/**
 * 主 Activity 为控制类,用于控制界面的显示
 */
public class MainActivity extends Activity {
    /**
     * Activity 组件的生命周期方法,在组件创建时调用,一般用于初始化信息
     */
    @Override
    protected void onCreate(Bundle savedInstanceState) {
        //调用父类方法,完成系统工作
        super.onCreate(savedInstanceState);
        //加载界面布局文件
        setContentView(R.layout.activity_main);
        //通过 id 获取界面控件 ListView
        ListView lv=(ListView) findViewById(R.id.lv);
        //创建数据源集合
        List<Map<String, Object>>data=new ArrayList<Map<String, Object>>();
        //创建集合中的第一个数据 Map
        Map<String, Object>map1=new HashMap<String, Object>();
        map1.put("nametext", "第一个功能");
        map1.put("iconid", R.drawable.ic_launcher);
        //创建集合中的第二个数据 Map
        Map<String, Object>map2=new HashMap<String, Object>();
        map2.put("nametext", "第二个功能");
        map2.put("iconid", R.drawable.ic_launcher);
        //创建集合中的第三个数据 Map
        Map<String, Object>map3=new HashMap<String, Object>();
        map3.put("nametext", "第三个功能");
        map3.put("iconid", R.drawable.ic_launcher);
        //创建集合中的第四个数据 Map
        Map<String, Object>map4=new HashMap<String, Object>();
        map4.put("nametext", "第四个功能");
        map4.put("iconid", R.drawable.ic_launcher);
        //创建集合中的第五个数据 Map
        Map<String, Object>map5=new HashMap<String, Object>();
        map5.put("nametext", "第五个功能");
        map5.put("iconid", R.drawable.ic_launcher);
        //将 Map 数据添加到集合中
        data.add(map1);
        data.add(map2);
```

```
        data.add(map3);
        data.add(map4);
        data.add(map5);
        //给 ListView 设置适配器 SimpleAdapter
        lv.setAdapter(new SimpleAdapter(this, data, R.layout.listitem,
                new String[] { "nametext", "iconid" }, new int[] { R.id.tv,
                        R.id.iv }));
    }

}
```

5. 案例运行效果

SimpleAdapter 案例的运行效果如图 5-3 所示。

图 5-3 SimpleAdapter 案例运行效果

5.2 ListView 列表控件的功能及使用

5.2.1 ListView 常用属性

在 Android 开发中 ListView 是比较常用的组件，它以列表的形式展示具体内容，并且能够根据数据的长度自适应显示，除此之外，ListView 还能处理用户的选择单击等操作。

1. ListView 的 XML 属性

ListView 的 XML 属性如表 5-1 所示。

表 5-1 ListView 的 XML 属性

XML 属性	XML 属性说明
android:divider	指定在列表条目之间显示的 drawable 或 color
android:dividerHeight	指定 divider 的高度
android:entries	构成 ListView 的数组资源的引用。对于某些固定的资源,这个属性提供了比在程序中添加资源更加简便的方式
android:footerDividersEnabled	当设为 false 时,ListView 将不会在各个 footer 之间绘制 divider。默认为 true
android:headerDividersEnabled	当设为 false 时,ListView 将不会在各个 header 之间绘制 divider。默认为 true

2. 继承自 AbsListView 的 XML 属性

AbsListView 是用于实现条目的虚拟列表的基类,ListView 继承自 AbsListView 的 XML 属性如表 5-2 所示。

表 5-2 ListView 继承自 AbsListView 的 XML 属性

XML 属性	XML 属性说明
android:cacheColorHint	表明这个列表的背景始终以单一、固定的颜色绘制,可以优化绘制过程
android:choiceMode	为视图指定选择的行为,可选的类型有 none、singleChoice、multipleChoice、multipleChoiceModal
android:drawSelectorOnTop	若设为 true,选择器将绘制在选中条目的上层。默认为 false
android:faseScrollEnabled	设置是否允许使用快速滚动滑块
android:listSelector	设置选中项显示的可绘制对象,可以是图片或者颜色属性
android:scrollingCache	设置在滚动时是否使用绘制缓存。若设为 true,则将使滚动表现更快速,但会占用更多内存。默认为 true
android:smoothScrollbar	为真时,列表会使用更精确的基于条目在屏幕上的可见像素高度的计算方法。默认该属性为真。如果适配器需要绘制可变高的条目,它应该设为 false。当该属性为 true 时,在适配器显示变高条目时,滚动条的把手会在滚动的过程中改变大小。当设为 false 时,列表只使用适配器中的条目数和屏幕上的可见条目来决定滚动条的属性
android:stackFromBottom	设置 GridView 和 ListView 是否将内容从底部开始显示
android:textFilterEnabled	当设为 true 时,列表会将结果过滤为用户类型。前提是这个列表的 Adapter 必须支持 Filterable 接口
android:transcriptMode	设置列表的 transcriptMode,有如下选项可选。 (1) disabled:禁用 TranscriptMode,也是默认值 (2) normal:当新条目添加进列表中并且已经准备好显示的时候,列表会自动滑动到底部以显示最新条目 (3) alwaysScroll:列表会自动滑动到底部,无论新条目是否已经准备好显示

3. 继承自 ViewGroup 的 XML 属性

ListView 继承自 ViewGroup 的 XML 属性如表 5-3 所示。

表 5-3　ListView 继承自 ViewGroup 的 XML 属性

XML 属性	XML 属性说明
android: addStatesFromChildren	设置这个 ViewGroup 的 drawable 状态是否包括子 View 的状态。若设为 true，当子 View 如 EditText 或 Button 获得焦点时，整个 ViewGroup 也会获得焦点
android:alwaysDrawnWithCache	设置 ViewGroup 在绘制子 View 时是否一直使用绘图缓存。默认为 true
android:animationCache	设置布局在绘制动画效果时是否为其子 View 创建绘图缓存。若设为 true，将会消耗更多的内存，要求持续时间更久的初始化过程，但表现更好。默认为 true
android:clipChildren	设置子 View 是否受限于在自己的边界内绘制。若设为 false，当子 View 所占用的空间大于边界时可以绘制在边界外。默认为 true
android:clipToPadding	定义布局间是否有间距。默认为 true
android: descendantFocusability	定义当寻找一个焦点 View 的时候，ViewGroup 与其子 View 之间的关系。可选项如下。 （1）beforeDescendants：ViewGroup 会比其子 View 更先获得焦点 （2）afterDescendants：只有当无子 View 想要获取焦点时，ViewGroup 才会获取焦点 （3）blockDescendants：ViewGroup 会阻止子 View 获取焦点
android:layoutAnimation	定义当 ViewGroup 第一次展开时的动画效果，也可人为地在第一次展开后调用

5.2.2　案例 ListView 具体使用

1. 案例功能描述

案例将实现在界面上以列表形式展示几条数据，数据内容较多，比较复杂。

2. 案例程序结构

案例中包括两个布局文件（activity_main.xml、listitem.xml），用于设计用户界面和显示 ListView 每个条目的布局；一个 Activity 组件（MainActivity 类），用于实现用户界面交互功能；一个实体类（Zhang），用于存储数据信息；一个自定义适配器类（ListAdapter），用于实现自定义的适配器。

3. 案例的实现步骤和思路

（1）创建 Android 项目。
（2）在 res 目录下的 layout 子目录中创建新的布局文件 activity_main.xml，最外层容器为 LinearLayout 布局,宽和高都是充满父容器,内部控件纵向摆放,创建一个内嵌 ListView,宽和高为充满包裹内容,id 为 listview。

(3) 在 res 目录下的 layout 子目录中创建新的布局文件 listview_item.xml,为 ListView 每个条目的布局,最外层容器为 LinearLayout 布局,宽和高都是充满父容器,内部控件纵向摆放。

① 创建第一个内嵌 TextView,宽和高为充满包裹内容,id 为 person_name,字号为 23sp;

② 创建第二个内嵌 TextView,宽和高为充满包裹内容,id 为 person_age;

③ 创建第三个内嵌 TextView,宽和高为充满包裹内容,id 为 person_email;

④ 创建第四个内嵌 TextView,宽和高为充满包裹内容,id 为 person_address。

(4) 编写 Zhang 类文件。

① 定义 4 个属性,用于封装界面的每个数据的 4 个信息;

② 定义全参的构造方法;

③ 给每个属性创建 get 和 set 方法;

④ 复写 toString 方法。

(5) 编写 ListAdapter 类文件,继承 ArrayAdapter,复写 getView 方法。

① 创建内部类 ViewHold,用于封装 4 个条目布局的控件;

② 定义 LayoutInflater 属性,在构造方法中实例化;

③ 定义构造方法,传入三个参数:应用程序上下文、条目布局资源和数据源数组;

④ 在 getView 方法中把数据和界面控件进行绑定。

(6) 编写 MainActivity 类文件,复写生命周期方法 onCreate。

① 定义属性,来实例化数据源数组;

② 在 onCreate 方法中,先调用父类的 onCreate 方法;

③ 使用 setContentView 方法加载布局文件 activity_main.xml;

④ 根据界面控件 id 使用 findViewById 方法,获取界面控件 ListView 并赋值给局部变量 listview;

⑤ 创建适配器,给 listview 设置适配器。

4. 案例参考代码

(1) 布局文件 activity_main.xml 代码如下。

```
<?xml version="1.0" encoding="utf-8"?>
<!--主界面布局,最外面的布局文件为线性布局 -->
<LinearLayout xmlns:android="http://schemas.android.com/apk/res/android"
    android:layout_width="fill_parent"
    android:layout_height="fill_parent"
    android:orientation="vertical">
    <ListView
        android:id="@+id/listview"
        android:layout_width="match_parent"
```

```
        android:layout_height="match_parent"
        ></ListView>
</LinearLayout>
```

（2）ListView 条目的布局文件 listitem.xml 的代码如下。

```
<?xml version="1.0" encoding="utf-8"?>
<!--ListView 每个条目的布局,最外面的布局文件为线性布局 -->
<LinearLayout
xmlns:android="http://schemas.android.com/apk/res/android"
android:id="@+id/linerlayout1"
android:orientation="vertical"
android:layout_height="fill_parent"
android:layout_width="fill_parent"
>
    <!--布局中第一个控件 TextView -->
    <TextView
        android:id="@+id/person_name"
        android:textSize="23sp"
        android:layout_width="wrap_content"
        android:layout_height="wrap_content"
    />
    <!--布局中第二个控件 TextView -->
    <TextView
        android:id="@+id/person_age"
        android:layout_width="wrap_content"
        android:layout_height="wrap_content"
    />
    <!--布局中第三个控件 TextView -->
    <TextView
        android:id="@+id/person_email"
        android:layout_width="wrap_content"
        android:layout_height="wrap_content"
    />
    <!--布局中第四个控件 TextView -->
    <TextView
        android:id="@+id/person_address"
        android:layout_width="wrap_content"
        android:layout_height="wrap_content"
    />
</LinearLayout>
```

（3）Zhang 类代码如下。

```
package com.example.android_demo5_3
```

```java
public class Zhang {
private String name;
    private int age;
    private String email;
    private String address;
    public String getName() {
        return name;
    }
    public int getAge() {
        return age;
    }
    public String getEmail() {
        return email;
    }
    public String getAddress() {
        return address;
    }
    public Zhang(String name, int age, String email, String address) {
        super();
        this.name=name;
        this.age=age;
        this.email=email;
        this.address=address;
    }
    @Override
    public String toString() {
        return "Person [name="+name+", age="+age+", email="+email
            +", address="+address+"]";
    }
}
```

（4）ListAdapter 类的代码如下。

```java
package com.example.android_demo5_3

import android.annotation.SuppressLint;
import android.content.Context;
import android.view.LayoutInflater;
import android.view.View;
import android.view.ViewGroup;
import android.widget.ArrayAdapter;
import android.widget.TextView;
/**
 * 实体类,封装数据信息
 */
```

```java
public class ListAdapter extends ArrayAdapter<Zhang>{
    //布局填充器
    private LayoutInflater mInflater;
    /**
     * 构造方法
     * @param context 应用程序上下文
     * @param textViewResourceId 条目布局文件资源
     * @param obj 数据源
     */
    public ListAdapter(Context context, int textViewResourceId,Zhang[] obj) {
        super(context, textViewResourceId,obj);
        this.mInflater=LayoutInflater.from(context);
    }
    /**
     * 返回条目布局
     */
    @SuppressLint("InflateParams") @SuppressWarnings("unused")
    @Override
    public View getView(int position, View convertView, ViewGroup parent) {
        //判断 converView 是否为空
        if(convertView==null){
            //创建新的 view 视图
            convertView=mInflater.inflate(R.layout.listview_item, null);
        }
        ViewHolder holder=null;
        //判断缓存对象是否为空
        if(holder==null){
            holder=new ViewHolder();
            //查找每个 ViewItem 中的各个子 View,放进 holder 中
            holder.name=(TextView) convertView.findViewById(R.id.person_name);
            holder.age= (TextView) convertView.findViewById(R.id.person_age);
            holder.email=(TextView) convertView.findViewById(R.id.person_email);
            holder.address= (TextView) convertView.findViewById(R.id.person_address);
            //保存对每个显示的 ViewItem 中, 各个子 View 的引用对象
            convertView.setTag(holder);
        }else{
            holder= (ViewHolder)convertView.getTag();
        }
        //获取当前要显示的数据
        Zhang person=getItem(position);
        holder.name.setText(person.getName());
        holder.age.setText(String.valueOf(person.getAge()));
        holder.email.setText(person.getEmail());
```

```java
        holder.address.setText(person.getAddress());
        return convertView;
    }
    /**
     * 内部类,缓存条目布局控件
     */
    private static class ViewHolder
    {
        TextView name;
        TextView age;
        TextView email;
        TextView address;
    }
}
```

(5) Activity 组件 MainActivity 类代码如下。

```java
package com.example.android_demo5_3

import android.app.Activity;
import android.os.Bundle;
import android.widget.ArrayAdapter;
import android.widget.ListView;

/**
 * 主 Activity 为控制类,用于控制界面的显示
 */
public class MainActivity extends Activity {
    //创建数据源
    private final static String[] data={"张飞","张辽","张角"};
    private Zhang[] data2=new Zhang[]{
        new Zhang("张飞",38,"zhangfei@gmail.com","燕山"),
        new Zhang("张辽",36,"zhangliao@sina.com","雁门"),
        new Zhang("张角",51,"zhangjiao@gmail.com","钜鹿")
    };
    /**
     * Activity 组件的生命周期方法,在组件创建时调用,一般用于初始化信息
     */
    @Override
    protected void onCreate(Bundle savedInstanceState) {
        //调用父类方法,完成系统工作
        super.onCreate(savedInstanceState);
        //加载界面布局文件
        setContentView(R.layout.activity_main);
        ListView listview= (ListView)findViewById(R.id.listview);
```

```
/*
*第一种:普通字符串
*/
ArrayAdapter<String>adapter1=new ArrayAdapter<String>(this,
        android.R.layout.simple_list_item_1,data);
/*
*第二种:自定义类对象
*/
ArrayAdapter<Zhang>adapter2=new ArrayAdapter<Zhang>(this,
        android.R.layout.simple_list_item_1,data2);
/*
*第三种:自定义适配器
*/
ListAdapter adapter3=new ListAdapter(this,R.layout.listview_item,data2);
listview.setAdapter(adapter3);
    }
}
```

5. 案例运行效果

ListView 案例的运行效果如图 5-4 所示。

5.2.3 响应单击事件

ListView 的响应单击事件是通过 OnItemClickListener 监听器来实现的,可以通过 setOnItemClickListener()方法为 ListView 注册条目单击事件进行监听与处理。例如:

图 5-4　ListView 案例运行效果

```
listview.setOnItemClickListener(new OnItemClickListener() {
    @Override
    public void onItemClick(AdapterView<?>arg0, View arg1, int arg2, long arg3)
    {
        //处理条目单击事件
    }
});
```

在上面代码的 onItemClick 方法中,参数含义如下。

(1) arg0 是指父 View,即 ListView;

(2) arg1 是当前 ListView 条目的 View,通过它可以获得该条目中的各个组件;

(3) arg2 是当前 ListView 条目的 id。这个 id 根据在适配器中的写法可以自己定义;

(4) arg3 是当前 ListView 条目在 ListView 中的相对位置。

5.3 GridView 网格控件的功能及使用

GridView 的使用与 ListView 的使用是类似的,默认不指定 GridView 一行显示的列数,GridView 就和 ListView 一样只显示一列,只有通过设置其 numColumns 属性才会在一行中显示多列,可以把 GridView 认为是一种特殊的 ListView。

5.3.1 GridView 常用属性

GridView 的常用 XML 属性如表 5-4 所示。

表 5-4 GridView 的常用 XML 属性

XML 属性	XML 属性含义	关联的方法
android:numColumns	GridView 的列数,如果设置 auto_fit,表示自动调整显示列数	setNumColumns(int)
android:columnWidth	每列的宽度,也就是 Item 的宽度	setColumnWidth(int)
android:verticalSpacing	两行之间的边距	setVerticalSpacing(int)
android:horizontalSpacing	两列之间的边距	setHorizontalSpacing(int)
android:stretchMode	缩放模式,如果设置为 columnWidth,表示缩放与列宽大小同步	setStretchMode(int)

5.3.2 案例 GridView 具体使用

1. 案例功能描述

案例将实现在界面上以表格形式展示几条数据,数据内容比较复杂,有图片和文字。

2. 案例程序结构

案例中包括两个布局文件(activity_main.xml、gridview_item.xml),用于设计用户界面和显示 GridView 每个条目的布局;一个 Activity 组件(MainActivity 类),用于实现用户界面交互功能。

3. 案例的实现步骤和思路

(1) 创建 Android 项目。

(2) 在 res 目录下的 layout 子目录中创建新的布局文件 activity_main.xml,最外层容器为 GridView 控件,宽和高都是充满父容器。

(3) 编写 MainActivity 文件,复写生命周期方法 onCreate。

① 在 onCreate 方法中,先调用父类的 onCreate 方法;

② 使用 setContentView 方法加载布局文件 activity_main.xml;

③ 根据界面控件 id 使用 findViewById 方法，获取界面控件 GridView 并赋值给局部变量 gridview；

④ 创建适配器，为 GridView 设置适配器。

4. 案例参考代码

（1）布局文件 activity_main.xml 的代码如下。

```xml
<?xml version="1.0" encoding="utf-8"?>
<!--主界面布局,最外面是 GridView 控件 -->
<GridView xmlns:android="http://schemas.android.com/apk/res/android"
android:id="@+id/gridview"
android:layout_width="fill_parent"
android:layout_height="fill_parent"
android:numColumns="auto_fit"
android:verticalSpacing="10dp"
android:horizontalSpacing="10dp"
android:columnWidth="90dp"
android:stretchMode="columnWidth"
android:gravity="center"
/>
```

（2）GridView 条目布局的布局文件 gridview_item 的代码如下。

```xml
<?xml version="1.0" encoding="utf-8"?>
<!--GridView 每个条目的布局,最外面的布局文件为相对布局 -->
<RelativeLayout xmlns:android="http://schemas.android.com/apk/res/android"
    android:layout_width="fill_parent"
    android:layout_height="wrap_content"
    android:paddingBottom="4dip">

    <!--布局中第一个控件 ImageView -->
    <ImageView
        android:id="@+id/ItemImage"
        android:layout_width="wrap_content"
        android:layout_height="wrap_content"
        android:layout_centerHorizontal="true">
    </ImageView>

    <!--布局中第二个控件 TextView -->
    <TextView
        android:id="@+id/ItemText"
        android:layout_width="wrap_content"
        android:layout_height="wrap_content"
        android:layout_below="@+id/ItemImage"
        android:layout_centerHorizontal="true"
        android:text="TextView01">
```

```
    </TextView>

</RelativeLayout>
```

(3) Activity 组件 MainActivity 类的代码如下。

```java
package com.example.android_demo5_4

import java.util.ArrayList;
import java.util.HashMap;

import android.app.Activity;
import android.os.Bundle;
import android.view.View;
import android.widget.AdapterView;
import android.widget.AdapterView.OnItemClickListener;
import android.widget.GridView;
import android.widget.SimpleAdapter;

/**
 * 主 Activity 为控制类,用于控制界面的显示
 */
public class MainActivity extends Activity {
    /**
     * Activity 组件的生命周期方法,在组件创建时调用,一般用于初始化信息
     */
    @Override
    protected void onCreate(Bundle savedInstanceState) {
        //调用父类方法,完成系统工作
        super.onCreate(savedInstanceState);
        //加载界面布局文件
        setContentView(R.layout.activity_main);
        //根据 id 获取界面控件 GridView
        GridView gridview= (GridView) findViewById(R.id.gridview);
        //生成动态数组,并且转入数据
        ArrayList<HashMap<String, Object>>lstImageItem=
            new ArrayList<HashMap<String, Object>>();
        for (int i=0; i<10; i++) {
            HashMap<String, Object>map=new HashMap<String, Object>();
            map.put("ItemImage", R.drawable.ic_launcher);    //添加图像资源的 ID
            map.put("ItemText", "NO."+String.valueOf(i));    //按序号做 ItemText
            lstImageItem.add(map);
        }
        //生成适配器的 ImageItem<=>动态数组的元素,两者一一对应
        SimpleAdapter saImageItems=new SimpleAdapter(this, lstImageItem,
            R.layout.gridview_item,
```

```
                new String[] { "ItemImage", "ItemText" },
                new int[] { R.id.ItemImage, R.id.ItemText });
        //添加适配器,显示数据
        gridview.setAdapter(saImageItems);
        //添加消息处理
        gridview.setOnItemClickListener(new ItemClickListener());
    }
    /**
     * 内部类,封装每个条目的单击监听器事件
     */
    class ItemClickListener implements OnItemClickListener {
        @SuppressWarnings("unchecked")
        public void onItemClick(AdapterView<?>arg0, View arg1, int arg2,
            long arg3) {
            //在本例中 arg2=arg3
            HashMap<String, Object>item= (HashMap<String, Object>) arg0
                .getItemAtPosition(arg2);
            //显示所选 Item 的 ItemText
            setTitle((String) item.get("ItemText"));
        }
    }
}
```

5. 案例运行效果

案例运行效果如图 5-5 和图 5-6 所示,图 5-5 显示的是没有单击 GridView 条目前的效果,图 5-6 显示的是单击 GridView 之后的效果,可以看到,单击条目后,条目显示的文本会在标题栏上显示出来。

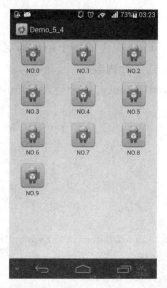

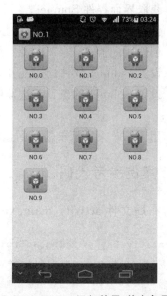

图 5-5　GridView 运行效果-没有单击条目前　　　　图 5-6　GridView 运行效果-单击条目后

5.4 Spinner 的功能及使用

Spinner 是 Android 中提供的一个下拉列表高级控件，当需要用户选择的时候，可以提供一个下拉列表将所有可选项列出来，用户每次只能选择所有项中的一项。Spinner 的选项来自于与之相关联的适配器中，它属于 AdapterView 类。下面将通过案例来说明 Spinner 的使用方法。

5.4.1 案例功能描述

案例将实现在界面上显示一个 Spinner 控件，默认显示文本为"周一"，单击 Spinner 控件后会弹出下拉选项，显示文本"周一"、"周二"至"周日"，单选条目后会弹出 Toast 提示 xxx＋选择文本。

5.4.2 案例程序结构

案例中包括一个布局文件(activity_main.xml)，用于设计用户界面；一个 Activity 组件(MainActivity 类)，用于实现用户界面交互功能。

5.4.3 案例的实现步骤和思路

(1) 创建 Android 项目。
(2) 编写布局文件 activity_main.xml，把 Spinner 控件显示到界面上。
(3) 实现 MainActivity 类，重写 onCreate 方法。
① 获取界面控件 Spinner；
② 创建数据源，用于 Spinner 控件的展开显示；
③ 创建适配器 ArrayAdapter，用于把数据源和 Spinner 控件的每个条目绑定上；
④ 设置 Spinner 控件的适配器，用于显示 Spinner 的展开内容；
⑤ 设置 Spinner 的条目单选监听器 OnItemSelectedListener，用于处理 Spinner 条目的单击；
⑥ 用 Toast 来显示单击后的显示内容。

5.4.4 案例参考代码

1. 布局文件 activity_main.xml 代码

```
<RelativeLayout xmlns:android=http://schemas.android.com/apk/res/android
    xmlns:tools=http://schemas.android.com/tools
    android:layout_width="match_parent"
    android:layout_height="match_parent"
```

```
        android:paddingBottom="@dimen/activity_vertical_margin"
        android:paddingLeft="@dimen/activity_horizontal_margin"
        android:paddingRight="@dimen/activity_horizontal_margin"
        android:paddingTop="@dimen/activity_vertical_margin"
        tools:context=".MainActivity">
    <Spinner android:id="@+id/spinner1"
        android:layout_width="match_parent"
        android:layout_height="wrap_content"
        android:layout_alignParentLeft="true"
        android:layout_alignParentTop="true"/>
</RelativeLayout>
```

2. Activity 组件 MainActivity 类代码

```java
package com.example.android_demo5_5;

import java.util.ArrayList;
import java.util.HashMap;
import java.util.List;
import java.util.Map;
import android.app.Activity;
import android.os.Bundle;
import android.view.Menu;
import android.view.View;
import android.widget.AdapterView;
import android.widget.ArrayAdapter;
import android.widget.SimpleAdapter;
import android.widget.Spinner;
import android.widget.Toast;

public class MainActivity extends Activity {
    //声明 Spinner 对象
    private Spinner spinner;
    @Override
    protected void onCreate(Bundle savedInstanceState) {
        super.onCreate(savedInstanceState);
        //加载界面
        setContentView(R.layout.activity_main);
        //获取界面的 Spinner 控件
        spinner=(Spinner) findViewById(R.id.spinner1);
        //使用数组作为数据源
        final String arr[]=new String[] { "周一", "周二", "周三", "周四", "周五",
            "周六", "周日" };
        //创建适配器对象,把数据源和界面控件 Spinner 中的每个条目进行绑定
```

```
    ArrayAdapter<String>arrayAdapter=new ArrayAdapter<String>(this,
        android.R.layout.simple_spinner_item, arr);
    //给 Spinner 设置上上面创建好的适配器对象
    spinner.setAdapter(arrayAdapter);
    //注册 Spinner 的单选事件,来处理 Spinner 的选项操作
    spinner.setOnItemSelectedListener(
        new AdapterView.OnItemSelectedListener() {
        @Override
        public void onItemSelected(AdapterView<?>parent, View view,
            int position,long id){
            Spinner spinner= (Spinner) parent;
            //使用 Toast 弹出提示信息
            Toast.makeText(getApplicationContext(),"xxxx"+
            spinner.getItemAtPosition(position),Toast.LENGTH_LONG).show();
        }
        @Override
        public void onNothingSelected(AdapterView<?>parent) {
            Toast.makeText(getApplicationContext(),
                "没有改变的处理",Toast.LENGTH_LONG).show();
        }
    });
}
```

5.4.5 案例运行效果

Spinner 案例的运行效果如图 5-7 所示。

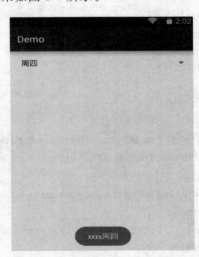

图 5-7 Spinner 案例运行效果图

5.5 菜单 Menu

菜单是用户界面中最常见的元素之一,使用非常频繁。在 Android 中,菜单被分为如下三种:选项菜单(OptionsMenu)、上下文菜单(ContextMenu)和子菜单(SubMenu)。其中最常用的就是选项菜单,在 Android SDK 2.x 版本,该菜单在单击 menu 按键后会在对应的 Activity 底部显示出来;Android SDK 3.x 和 4.x 版本的新版本中,菜单一般显示在 ActionBar 上。本节主要讨论选项菜单的定义和使用方法。

5.5.1 使用 xml 定义 Menu

菜单资源文件必须放在项目的 res/menu 目录中。定义菜单的 XML 必须使用<menu>标签作为根结点,另外两个标签是<item>和<group>,用于设置菜单项和分组。<menu>标签没有任何属性,但可以嵌套在<item>标签中,表示子菜单的形式。不过<item>标签中不能再嵌入<item>标签。

1. <item>标签属性

(1) id:表示菜单项的资源 ID。

(2) menuCategory:同种菜单项的种类。该属性可取 4 个值:container、system、secondary 和 alternative。通过 menuCategroy 属性可以控制菜单项的位置。例如将属性设为 system,表示该菜单项是系统菜单,应放在其他种类菜单项的后面。

(3) orderInCategor:同种类菜单的排列顺序。该属性需要设置一个整数值。例如 menuCategory 属性值都为 system 的三个菜单项(item1、item2 和 item3)。将这三个菜单项的 orderInCategory 属性值设为 3、2、1,那么 item3 会显示在最前面,而 item1 会显示在最后面。

(4) title:菜单项标题,即菜单项显示的文本。

(5) titleCondensed:菜单项的短标题。当菜单项标题太长时会显示该属性值。

(6) icon:菜单项图标资源 ID。

(7) alphabeticShortcut:菜单项的字母快捷键。

(8) numericShortcut:菜单项的数字快捷键。

(9) checkable:表示菜单项是否带复选框。该属性可设计为 true 或 false。

(10) checked:如果菜单项带复选框(checkable 属性为 true),该属性表示复选框默认状态是否被选中。可设置的值为 true 或 false。

(11) visible:菜单项默认状态是否可视。

(12) enable:菜单项默认状态是否被激活。

2. <group>标签属性

(1) id:表示菜单组的 ID。

（2）menuCategory：与＜item＞标签的同名属性含义相同，只是作用域为菜单组。

（3）orderInCategory：与＜item＞标签的同名属性含义相同，只是作用域为菜单组。

（4）checkableBehavior：设置该组所有菜单项上显示的选择组件（CheckBox 或 Radio Button）。如果将该属性值设为 all，显示 CheckBox 组件；如果设为 single，显示 Radio Button 组件；如果设为 none，显示正常的菜单项（不显示任何选择组件）。

要注意的是，Android SDK 官方文档在解释该属性时有一个笔误，原文是：

Whether the items are checkable. Valid values：none，all（exclusive/radiobuttons），single（non-exclusive/checkboxes）.

正确的应该是：

all（non-exclusive/checkboxes），single（exclusive/radiobuttons）.

（5）visible：表示当前组中所有菜单项是否显示，该属性可设置的值是 true 或 false。

（6）enable：表示当前组中所有菜单项是否被激活，该属性可设置的值是 true 或 false。

5.5.2 使用代码定义 Menu

可以在 Activity 的 onCreateOptionsMenu(Menu menu)方法或者 onPrepareOptionsMenu(Menu menu)方法中，利用菜单对象 Menu 的 add()方法来添加菜单项，add()方法原型为：

```
menu.add ((int groupId, int itemId, int order, charsequence title) .setIcon (drawable ID)
```

上面的 4 个参数，依次介绍如下。

（1）groupId：组别，如果不分组的话就写 Menu.NONE；

（2）itemId：Android 根据这个 Id 来确定不同的菜单；

（3）order：哪个菜单项在前面由这个参数的大小决定；

（4）title：菜单项的显示文本。

add()方法返回的是 MenuItem 对象，调用其 setIcon()方法，可以为相应的菜单项 MenuItem 设置图标。例如：

```
public boolean onCreateOptionsMenu(Menu menu) {
    //用代码构建菜单
    menu.add(Menu.NONE, Menu.NONE, 1, "菜单 1");
    menu.add(Menu.NONE, Menu.NONE, 2, "菜单 2");
    menu.add(Menu.NONE, Menu.NONE, 3, "菜单 3");
    menu.add(Menu.NONE, Menu.NONE, 4, "菜单 4");
    menu.add(Menu.NONE, Menu.NONE, 5, "菜单 5");
    menu.add(Menu.NONE, Menu.NONE, 6, "菜单 6");
```

```
        return true;
}
```
上面代码的运行效果如图 5-8 所示。

5.5.3 使用菜单

在 Activity 里面，一般通过以下函数来使用选项菜单。

（1）Activity：：onCreateOptionsMenu（Menu menu）：创建选项菜单，这个函数只会在菜单第一次显示时调用。

（2）Activity：：onPrepareOptionsMenu（Menu menu）：更新改变选项菜单的内容，这个函数会在菜单每次显示时调用。

（3）Activity：：onOptionsItemSelected（MenuItem item）：处理选中的菜单项，重写 onOptionsItemSelected（MenuItem item）这个方法就可以做响应操作了。

图 5-8 使用代码创建 Menu

5.6 案例菜单 Menu

5.6.1 案例功能描述

案例将实现在 ActionBar 上显示主菜单，单击后会弹出带有图片和文字的菜单项，再次单击菜单项会显示提示信息。

5.6.2 案例程序结构

案例中包括一个 Activity 布局文件 activity_main.xml；一个菜单布局文件（main.xml）；一个 Activity 组件（MainActivity 类），用于实现用户界面交互功能。

5.6.3 案例的实现步骤和思路

（1）创建 Android 项目。
（2）在 res 目录下的 layout 子目录中创建新的布局文件 activity_main.xml。
（3）在 res 目录下的 menu 子目录中创建新的菜单布局文件 main.xml，有三个子菜单项。
（4）编写 MainActivity 文件。
① 复写生命周期方法 onCreate，使用 setContentView 方法加载布局文件；
② 复写 onCreateOptionsMenu 方法来创建菜单，在方法中加载菜单布局；
③ 复写 onOptionsItemSelected 方法，根据菜单 id 来处理不同菜单项的单击。

5.6.4 案例参考代码

1. 菜单布局文件 main.xml 代码

```xml
<!--菜单布局文件 -->
<menu xmlns:android="http://schemas.android.com/apk/res/android">
    <!--第一菜单项 -->
    <item
        android:id="@+id/menu_settings"
        android:orderInCategory="1"
        android:icon="@drawable/setting"
        android:title="@string/menu_setting"/>
    <!--第二菜单项 -->
    <item
        android:id="@+id/menu_about"
        android:orderInCategory="2"
        android:icon="@drawable/about"
        android:title="@string/menu_about"/>
    <!--第三菜单项 -->
    <item
        android:id="@+id/menu_quit"
        android:orderInCategory="3"
        android:icon="@drawable/quit"
        android:title="@string/menu_quit"/>
</menu>
```

2. Activity 组件 MainActivity 类代码

```java
package com.example.android_demo5_6

import android.app.Activity;
import android.os.Bundle;
import android.view.Menu;
import android.view.MenuItem;
import android.view.SubMenu;
import android.widget.Toast;

/**
 * 主 Activity 为控制类,用于控制界面的显示
 */
public class MainActivity extends Activity {
    /**
     * Activity组件的生命周期方法,在组件创建时调用,一般用于初始化信息
```

```java
     */
    @Override
    protected void onCreate(Bundle savedInstanceState) {
        super.onCreate(savedInstanceState);
        setContentView(R.layout.activity_main);
    }
    /**
     * 复写创建菜单的方法
     */
    @Override
    public boolean onCreateOptionsMenu(Menu menu) {
        //在 ActionBar 添加菜单
        SubMenu subMenu=menu.addSubMenu("");
        //添加菜单布局
        getMenuInflater().inflate(R.menu.main, subMenu);
        //得到刚添加的菜单
        MenuItem item=subMenu.getItem();
        //设置 ActionBar 上的菜单的标题、按钮及显示选项
        item.setTitle("菜单");
        item.setIcon(R.drawable.menusub);
        item.setShowAsAction(MenuItem.SHOW_AS_ACTION_ALWAYS |
            MenuItem.SHOW_AS_ACTION_WITH_TEXT);
        return super.onCreateOptionsMenu(menu);
    }
    /**
     * 复写菜单单击事件的处理方法
     */
    @Override
    public boolean onOptionsItemSelected(MenuItem item) {
        //根据不同的菜单项 id 来处理不同的菜单项单击
        switch(item.getItemId()){
            case R.id.menu_about:
                Toast.makeText(MainActivity.this, ""+"关于",
                    Toast.LENGTH_SHORT).show();
                break;
            case R.id.menu_settings:
                Toast.makeText(MainActivity.this, ""+"设置",
                    Toast.LENGTH_SHORT).show();
                break;
            case R.id.menu_quit:
                Toast.makeText(MainActivity.this, ""+"退出",
                    Toast.LENGTH_SHORT).show();
                break;
            default:
```

```
            break;
        }
        return super.onOptionsItemSelected(item);
    }
}
```

5.6.5 案例运行效果

案例运行效果如图 5-9 和图 5-10 所示。

图 5-9　案例运行效果-单击菜单

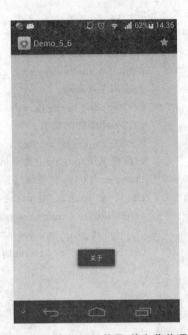

图 5-10　案例运行效果-单击菜单项

5.7　对话框 Dialog

在 Android 中对话框是模态异步的，弹出对话框与用户交互时后台执行线程还是接着执行。

Android 提供了丰富的对话框支持，它提供了如下 4 种常用的对话框。

（1）AlertDialog：警告对话框，使用最广泛、功能最丰富的一个对话框。
（2）ProgressDialog：进度条对话框。
（3）DatePickerDialog：日期对话框。
（4）TimePickerDialog：时间对话框。

所有的对话框都是直接或间接继承自 Dialog 类，而 AlterDialog 直接继承自 Dialog，其他的几个类均继承自 AlterDialog。

对于 Android 内置的 AlterDialog，它可以包含一个标题、一个内容消息或者一个选择列表,最多支持三个按钮。如果要创建 AlterDialog,推荐使用它的一个内部类 AlterDialog.Builder 创建。使用 Builder 对象,可以设置 AlterDialog 的各种属性,最后通过 Builder.create()就可以得到 AlterDialog 对象。一般使用 Builder.show()方法显示对话框,它会返回一个 AlterDialog 对象。

5.7.1 简单对话框

直接设置 AlterDialog 的一些属性,就可以显示提示信息,主要方法如下。
(1) AlterDialog create():创建一个 AlterDialog。
(2) AlterDialog show():将 AlterDialog 显示在屏幕上。
(3) AlterDialog.Builder setTitle():设置标题。
(4) AlterDialog.Builder setIcon():设置标题的图标。
(5) AlterDialog.Builder setMessage():设置标题的内容。
(6) AlterDialog.Builder setCancelable():设置是否模态,一般设置为 false,表示模态,要求用户必须采取动作才能继续进行剩下的操作。

当一个对话框调用了 show()方法后,会展示到屏幕上;如果需要消除它,可以使用 cancel()和 dismiss()使对话框取消或者消除,这两个方法的作用是一样的,不过推荐使用 dismiss()。

通过下面的代码可以显示一个最简单的 AlterDialog。

```
//显示简单对话框
public void showNormalDialog() {
    AlertDialog.Builder builder=new
        AlertDialog.Builder(MainActivity.this);
    builder.setTitle("提示");
    builder.setMessage("这是一个简单对话框");
    builder.setIcon(R.drawable.ic_launcher);
    builder.setCancelable(false);
    builder.setPositiveButton("确定",
        new DialogInterface.OnClickListener() {
            @Override
            public void onClick(DialogInterface dialog, int which) {
                dialog.cancel();
            }
        });
    builder.create().show();
}
```

上面代码的运行效果如图 5-11 所示。

5.7.2 多按钮对话框

AlterDialog 内置了三个按钮,可以直接使用

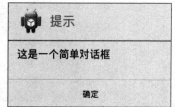

图 5-11 简单对话框

setXxxButton()方法进行设置,下面是这三个方法:

(1) AlterDialog. Builder setPositiveButton(CharSquence text, DialogInterFace. OnClickListener)一般用于OK、"确定"或者"继续"等操作。

(2) AlterDialog. Builder setNegativeButton(CharSquence text, DialogInterFace. OnClickListener)一般用于"取消"操作。

(3) AlterDialog. Builder setNeutralButton(CharSquence text, DialogInterFace. OnClickListener)一般用于"忽略"、"以后提醒我"等操作。

上面介绍的 DialogInterface 接口,还提供了一系列的事件响应,这三个按钮均需要传递一个 DialogInterFace.OnClickListener 接口对象,实现其单击事件的触发。在这个接口中需要实现 onClick(DialogInterface dialog,int which),其中 dialog 为当前触发事件的对话框对象接口,可以直接强制转换为 AlterDialog 进行操作;which 为单击按钮的标识符,是一个整型数据。对于这三个按钮而言,每个按钮使用不同的 int 类型数据进行标识,即 Positive(-1)、Negative(-2)、Neutral(-3)。

除了专门为按钮单击实现的 DialogInterFace.OnClickListener 事件外,DialogInterface 还提供了一些其他的事件,供 Dialog 对象响应,这些事件只是对 Dialog 声明周期各个状态的响应,介绍如下。

(1) interface DialogInterface.OnCancelListener:当对话框调用 cancel()方法的时候触发。

(2) interface DialogInterface.OnDismissListener:当对话框调用 dismiss()方法的时候触发。

(3) interface DialogInterface.OnShowListener:当对话框调用 show()方法的时候触发。

(4) interface DialogInterface.OnMultiChoiceListener:当对话框使用多选列表,并且选中的时候触发。

下面的代码将展示如何显示一个多按钮对话框,并对按钮事件进行监听和响应。

```
//显示多选按钮普通对话框
    private void showMoreButtonNormalDialog() {
        AlertDialog.Builder builder=new
            AlertDialog.Builder(MainActivity.this);
        builder.setTitle("提示");
        builder.setMessage("这是一个包含三个按钮的对话框");
        builder.setIcon(R.drawable.ic_launcher);
        builder.setPositiveButton("确定",
            new DialogInterface.OnClickListener() {
            @Override
            public void onClick(DialogInterface dialog, int which) {
                Toast.makeText(MainActivity.this,
                    "单击了确定按钮", Toast.LENGTH_SHORT).show();
                dialog.dismiss();
```

```
            }
        });
        builder.setNegativeButton("取消",
            new DialogInterface.OnClickListener() {
            @Override
            public void onClick(DialogInterface dialog, int which) {
                Toast.makeText(MainActivity.this,
                    "单击了取消按钮", Toast.LENGTH_SHORT).show();
                dialog.dismiss();
            }
        });
        builder.setNeutralButton("忽略",
            new DialogInterface.OnClickListener() {
            @Override
            public void onClick(DialogInterface dialog, int which) {
                Toast.makeText(MainActivity.this,
                    "单击了忽略按钮", Toast.LENGTH_SHORT).show();
                dialog.cancel();
            }
        });
        builder.show();
    }
```

上面代码的运行效果如图5-12所示。

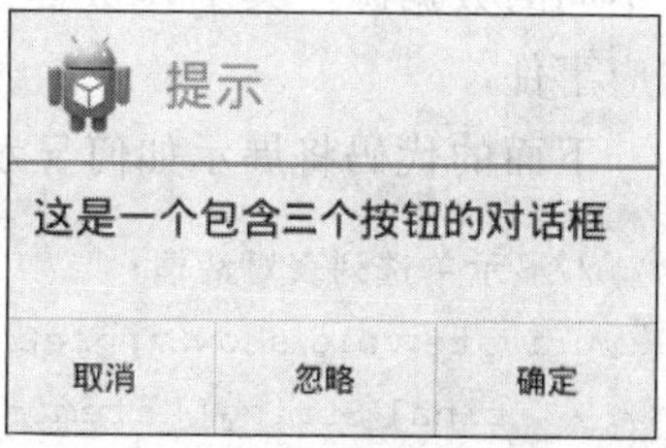

图 5-12 多按钮对话框

5.7.3 列表对话框

AlterDialog 除了展示一些提示信息，还可以展示一种列表的形式，需要使用到 Builder. setItems(CharSequence[] items, DialogInterface. OnClickListener listener) 方法进行设置，它需要传递一个 CharSequenece 类型的数组，以绑定列表数据，它同样需要传递一个 DialogInterface. OnClickListener 接口，以响应列表项的单击。在这个接口中 onClick 方法的 which 参数，为当前单击触发项的条目中的下标。

下面的代码将展示如何显示一个列表对话框，当用户单击某一列表项时，会显示提示信息。

```
//显示列表对话框
private void showListDialog() {
    final String[] items=new String[] { "读书","听音乐","爬山" };
    AlertDialog.Builder builder=
        new AlertDialog.Builder(MainActivity.this);
    builder.setTitle("请选择你的兴趣、爱好");
    //items 使用全局的 finalCharSequenece 数组声明
    builder.setItems(items, new DialogInterface.OnClickListener() {
        @Override
```

```java
    public void onClick(DialogInterface dialog, int which) {
        //TODO Auto-generated method stub
        String select_item=items[which].toString();
        Toast.makeText(MainActivity.this,
            "你选择了<"+select_item+">",Toast.LENGTH_SHORT).show();
    }
});
builder.show();
}
```

上面代码的运行效果如图 5-13 所示。

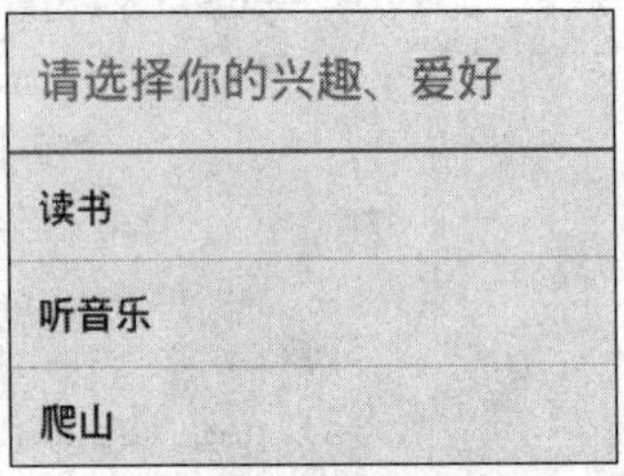

5.7.4 单选列表对话框

图 5-13 列表对话框

AlterDialog 还可以使用一种单选的列表样式,使用 Builder.setSingleChoiceItems（CharSequenece [ ] items, int checkedItem, DialogInterface.OnClickListener listener）方法,这个方法具有多项重载,主要是为了应对不同的数据源。其中,items 为列表项数组,checkedItem 为初始选项,listener 为单击响应事件。

下面的代码将展示如何显示一个单选列表对话框。

```java
//显示单选列表对话框
private void showSingleChoiceDialog() {
    final String[] items=new String[] { "读书", "听音乐", "爬山" };
    AlertDialog.Builder builder=
        new AlertDialog.Builder(MainActivity.this);
    builder.setTitle("请选择你的兴趣、爱好");
    builder.setSingleChoiceItems(items, 1,
        new DialogInterface.OnClickListener() {
            @Override
            public void onClick(DialogInterface dialog, int which) {
                //TODO Auto-generated method stub
                String select_item=items[which].toString();
                Toast.makeText(MainActivity.this,
                    "你选择了<"+select_item+">",
                    Toast.LENGTH_SHORT)
                    .show();
            }
        });
    builder.setPositiveButton("确定",
        new DialogInterface.OnClickListener() {
            @Override
            public void onClick(DialogInterface dialog, int which) {
                dialog.dismiss();
```

            }
        });
        builder.show();
}
```

上面代码的运行效果如图 5-14 所示。

### 5.7.5 复选列表对话框

图 5-14 单选列表对话框

AlterDialog 除了单选列表，还有复选（多选）列表。可以使用 Builder. setMulti ChoiceItems (CharSequence[] items, boolean[] checkedItems, DialogInterface. OnMultiChoiceClickListener listener) 方法，这个方法也同样具有多个重载。其中，items 以一个数组为数据源，checkedItems 是默认选项，listener 为多选项单击触发事件。

下面代码将展示如何显示一个复选列表对话框。

```java
//显示复选列表对话框
private void showMultiChoiceDialog() {
    final String[] items=new String[] { "读书", "听音乐", "爬山" };
    AlertDialog.Builder builder=
        new AlertDialog.Builder(MainActivity.this);
    builder.setTitle("请选择你的兴趣、爱好");
    builder.setMultiChoiceItems(items, new boolean[] { true, false, true },
        new DialogInterface.OnMultiChoiceClickListener() {
            @Override
            public void onClick(DialogInterface dialog, int which,
                boolean isChecked) {
                //TODO Auto-generated method stub
                String select_item=items[which].toString();
                Toast.makeText(MainActivity.this,
                    "你选择了<"+select_item+">",
                    Toast.LENGTH_SHORT)
                    .show();
            }
        });
    builder.setPositiveButton("确定",
        new DialogInterface.OnClickListener() {
        @Override
        public void onClick(DialogInterface dialog, int which) {
            dialog.dismiss();
        }
    });
    builder.show();
}
```
```

上面代码的运行效果如图 5-15 所示。

5.7.6 自定义对话框

可以使用自定义样式的 XML 布局文件,作为 AlertDialog 的样式展示在屏幕上。对于定制的 XML 文件,可以使用 LayoutInflater. from(Context). inflate(int,ViewGroup)的方式对其进行动态加载,然后使用 Builder. setView(View)把加载的视图与 Builder 对象进行关联,最后调用 show()方法显示即可。

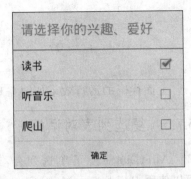

图 5-15 复选列表对话框

下面的代码将展示如何显示一个自定义对话框。

1. 自定义对话框布局文件 my_dialog. xml

```xml
<?xml version="1.0" encoding="utf-8"?>
<!--自定义登录对话框布局,最外面是线性布局 -->
<LinearLayout xmlns:android="http://schemas.android.com/apk/res/android"
    android:layout_width="match_parent"
    android:layout_height="match_parent"
    android:orientation="vertical">

    <!--布局中第一个控件 EditText -->
    <EditText
        android:id="@+id/editText1"
        android:layout_width="match_parent"
        android:layout_height="wrap_content"
        android:ems="10"
        android:hint="用户名">
        <requestFocus />
    </EditText>

    <!--布局中第二个控件 EditText -->
    <EditText
        android:id="@+id/editText2"
        android:layout_width="match_parent"
        android:layout_height="wrap_content"
        android:ems="10"
        android:hint="密    码" />

    <!--布局中第三个控件 Button -->
    <Button
```

```
        android:id="@+id/btnCustom"
        android:layout_width="match_parent"
        android:layout_height="wrap_content"
        android:text="登录" />

</LinearLayout>
```

2. 自定义对话框定义代码

```
//显示自定义对话框
AlertDialog alertDialog=null;
private void showSelfDialog() {
    AlertDialog.Builder builder=
        new AlertDialog.Builder(MainActivity.this);
    View view=LayoutInflater.from(MainActivity.this).inflate(
        R.layout.dialog_signin, null);
    Button btn=(Button) view.findViewById(R.id.btnCustom);
    btn.setOnClickListener(new View.OnClickListener() {
        @Override
        public void onClick(View v) {
            alertDialog.dismiss();
            Toast.makeText(MainActivity.this, "登录验证",
                Toast.LENGTH_SHORT).show();
        }
    });
    builder.setView(view);
    alertDialog=builder.show();
}
```

上面代码的运行效果如图 5-16 所示。

图 5-16 自定义对话框

5.7.7 进度对话框

可以使用进度对话框 ProgressDialog 来显示一个进度信息,提示用户等待。ProgressDialog 的使用方式大部分可以参考 ProgressBar,ProgressDialog 其实就是一个封装了 ProgressBar 的对话框。

ProgressDialog 有两种显示方式,一种是以一个滚动的环状图标,可以显示一个标题和一段文本内容的等待对话框;另外一种是带刻度的进度条,和常规的进度条用法一致。两种样式通过 ProgressDialog.setProgressStyle(int style)设置,ProgressDialog 的两个常量为:STYLE_HORIZONTAL——刻度滚动;STYLE_SPINNER——图标滚动,是默认选项。

对于图标滚动,可以使用两种方式实现,一种是调用构造函数,再设置对应的属性;

另外一种是直接使用 ProgressDialog 的静态方法 show(),直接返回一个 ProgressDialog 对象,再调用 show()方法显示。

下面的代码将展示如何显示一个进度对话框。

```
//显示进度对话框
private void showPregressDialog() {
    ProgressDialog progressDialog;
    progressDialog=new ProgressDialog(MainActivity.this);
    progressDialog.setIcon(R.drawable.ic_launcher);
    progressDialog.setTitle("请稍等");
    progressDialog.setMessage("正在加载…");
    progressDialog.show();
}
```

上面代码的运行效果如图 5-17 所示。

图 5-17　进度对话框

5.7.8　自定义进度对话框

对于有刻度的 ProgressDialog,除了从 AlertDialog 中继承来的属性,有一些必要的属性需要设置,介绍如下。

(1) setMax(int max):最大刻度。

(2) setProgress(int value):第一进度。

(3) setSecondaryProgress(int value):第二进度。

下面的代码将展示如何显示一个自定义进度对话框。

```
//显示自定义进度对话框
ProgressDialog progressDialog;
private void showCustomPregressDialog() {
    progressDialog=new ProgressDialog(MainActivity.this);
    progressDialog.setIcon(R.drawable.ic_launcher);
    progressDialog.setTitle("请稍等");
    progressDialog.setMessage("正在加载…");
    progressDialog.setProgressStyle(ProgressDialog.STYLE_HORIZONTAL);
    progressDialog.show();
    new Thread(new Runnable() {
        @Override
        public void run() {
            try {
                for (int i=0; i<=100; i++) {
                    Thread.sleep(100);
                    progressDialog.setProgress(i);
                }
            } catch (Exception e) {
```

```
                e.printStackTrace();
            } finally {
                progressDialog.dismiss();
            }
        }
    }).start();
}
```

图 5-18 自定义进度对话框

上面代码的运行效果如图 5-18 所示。

5.8 用 Fragment 分割用户界面

Android 3.0 中引入了 Fragment(片段、碎片)的概念,其目的是为了解决不同屏幕分辨率的动态和灵活的 UI 设计。可以把 Fragment 当成 Activity 的一个界面的一个组成部分,甚至 Activity 的界面可以完全由不同的 Fragment 组成。通过将 Activity 的布局分散到 Fragment 中,可以在运行时修改 Activity 的外观。当一个 Fragment 指定了自身的布局时,它能和其他 Fragment 配置成不同的组合,在 Activity 中为不同的屏幕尺寸修改布局配置(小屏幕可能每次显示一个 Activity,而大屏幕则可以显示两个或更多的 Fragment)。Fragment 必须被写成可重用的模块,因为 Fragment 拥有自己的布局和自己的生命周期,可以接收、处理用户事件。更为重要的是,可以动态地添加、替换和移除某个 Fragment,这对于让界面在不同的屏幕尺寸下都能给用户完美的体验尤其重要。

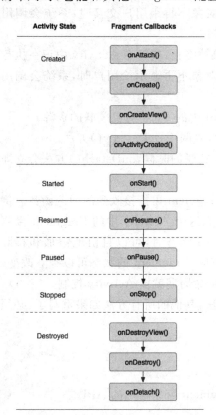

图 5-19　Activity 和 Fragment 的生命周期间的关系

5.8.1 Fragment 的生命周期

Fragment 不能独立存在,它必须嵌入到 Activity 中,而且 Fragment 的生命周期直接受所在的 Activity 的影响。例如,当 Activity 暂停时,它拥有的所有的 Fragment 都暂停了;当 Activity 销毁时,它拥有的所有 Fragment 都被销毁。然而,当 Activity 运行时(在 onResume()之后、onPause()之前),可以单独地操作每个 Fragment,例如添加或删除它们。执行上述针对 Fragment 的事务时,可以将事务添加到一个栈中,这个栈被 Activity 管理,栈中的每一条都是一个 Fragment 的一次事务。有了这个栈,就可以反向执行 Fragment 的事务,这样就可以在 Fragment 级支持"返回"键。图 5-19 说明了 Activity 和

Fragment 的生命周期间的关系。

1. 显示 Fragment 时（跟用户交互）要调用的核心的生命周期方法

（1）把 Fragment 对象跟 Activity 关联时，调用 onAttach(Activity)方法；

（2）Fragment 对象初始创建时，调用 onCreate(Bundle)方法；

（3）onCreateView(LayoutInflater,ViewGroup,Bundle)方法用于创建和返回跟 Fragment 关联的 View 对象；

（4）onActivityCreate(Bundle)方法会告诉 Fragment 对象，它所依附的 Activity 对象已经完成了 Activity.onCreate()方法的执行；

（5）onStart()方法会让 Fragment 对象显示给用户（在包含该 Fragment 对象的 Activity 被启动后）；

（6）onResume()会让 Fragment 对象跟用户交互（在包含该 Fragment 对象的 Activity 被启动恢复后）。

2. Fragment 对象不再使用时，要反向回调的方法

（1）因为 Fragment 对象所依附的 Activity 对象被挂起，或者在 Activity 中正在执行一个修改 Fragment 对象的操作，而导致 Fragment 对象不再跟用户交互时，系统会调用 Fragment 对象的 onPause()方法。

（2）因为 Fragment 对象所依附的 Activity 对象被终止，或者在 Activity 中正在执行一个修改 Fragment 对象的操作，而导致 Fragment 对象不再显示给用户时，系统会调用 Fragment 对象的 onStop()方法。

（3）onDestroyView()方法用于清除与 Fragment 中的 View 对象关联的资源。

（4）Fragment 对象的状态被最终清理完成之后，要调用 onDestroy()方法。

（5）在 Fragment 对象不再与它依附的 Activity 关联的时候，onDetach()方法会立即被调用。

从图 5-18 中可以看出 Activity 的状态决定了 Fragment 可能接收到的回调函数。例如，当 Activity 接收到它的 onCreate()回调函数，那么这个 Activity 中的 Fragment 接收到了 onActivityCreated()。当 Activity 处于 Resumed 状态时，可以自由地添加和移除 Fragment，也就是说，只有 Activity 在 Resumed 状态时，Fragment 的状态可以独立改变。但是，当 Activity 离开 Resumed 状态，Fragment 的生命周期就被 Activity 控制。

需要注意：除了 Fragment 的 onCreateView 方法，其他的所有方法如果重写了，必须调用父类对于该方法的实现。

5.8.2 设计基于 Fragment 的应用

要想设计基于 Fragment 的应用，需要了解 Fragment 家族常用的 API。

1. Fragment 常用的 API

Fragment 常用的三个类如下。

(1) android.app.Fragment：主要用于定义 Fragment。

(2) android.app.FragmentManager：主要用于在 Activity 中操作 Fragment。通过调用 Activity 的 getFragmentManager()方法可以取得 FragmentManage 的实例。

(3) android.app.FragmentTransaction：对 Fragment 进行添加、移除、替换及执行其他动作。

在使用 FragmentTransaction 的方法前，首先需要取得 FragmentManage 的实例，再利用 FragmentManage 的 beginTransaction() 方法开启一个事务，获取一个 FragmentTransaction 对象。例如：

```
FragmentManager fragmentManager=getFragmentManager();
FragmentTransaction fragmentTransaction = fragmentManager.beginTransaction
();
```

FragmentTransaction 的方法主要如下几个。

(1) FragmentTransaction.add()

往 Activity 中添加一个 Fragment。

(2) FragmentTransaction.remove()

从 Activity 中移除一个 Fragment，如果被移除的 Fragment 没有添加到回退栈，这个 Fragment 实例将会被销毁。回退栈（Back Stack）由 Activity 管理，允许用户通过按 BACK 键返回到前一个 Fragment 状态。

(3) FragmentTransaction.replace()

使用另一个 Fragment 替换当前的 Fragment。

(4) FragmentTransaction.hide()

隐藏当前的 Fragment，仅仅是设为不可见，并不会销毁。

(5) FragmentTransaction.show()

显示之前隐藏的 Fragment。

(6) FragmentTransaction.detach()

会将 View 从 UI 中移除，和 remove() 不同，此时 Fragment 的状态依然由 FragmentManager 维护。

(7) FragmentTransaction.attach()

重建 View 视图，附加到 UI 上并显示。

(8) FragmentTransaction.commit()

提交一个事务。在一个事务开启到提交过程中可以进行多个 Fragment 的添加、移除和替换等操作。需要注意，FragmentTransaction 的 commit()方法一定要在 Activity.onSaveInstance()方法之前调用。

了解了 Fragment 常用的 API 之后，接下来将介绍如何使用 Fragment。通常使用 Fragment 有两种方式：静态使用和动态使用。

2. 静态使用 Fragment

这是使用 Fragment 最简单的一种方式，把 Fragment 当成普通的控件，直接写在

Activity 的布局文件中,其主要步骤如下。

(1) 继承 Fragment,重写 onCreateView()方法,决定 Fragement 的布局;

(2) 在 Activity 中声明此 Fragment,就和普通的 View 一样。

下面将通过一个实例说明 Fragment 的静态使用方法,实例中使用两个 Fragment 作为 Activity 的布局,一个 Fragment 用于标题布局(titlefragment.xml 为布局文件,TitleFragment.java 为 Fragment 类文件),另一个 Fragment 用于内容布局(contentfragment.xml 为布局文件,ContentFragment.java 为 Fragment 类文件);实例中还包含一个 Activity 类文件 MainActivity.java 及其布局文件 activity_main.xml。

(1) titlefragment.xml

```xml
<?xml version="1.0" encoding="utf-8"?>
<RelativeLayout xmlns:android="http://schemas.android.com/apk/res/android"
    android:layout_width="match_parent"
    android:layout_height="45dp"
    android:background="@drawable/bg">

    <ImageButton
        android:id="@+id/id_title_left_btn"
        android:layout_width="wrap_content"
        android:layout_height="wrap_content"
        android:layout_centerVertical="true"
        android:layout_marginLeft="3dp"
        android:background="@drawable/about" />

    <TextView
        android:layout_width="fill_parent"
        android:layout_height="fill_parent"
        android:gravity="center"
        android:text="我是TitleFragment"
        android:textColor="#000"
        android:textSize="20sp"
        android:textStyle="bold" />

</RelativeLayout>
```

(2) TitleFragment 类文件

```java
package com.example.android_demo5_8;
import android.app.Fragment;
import android.os.Bundle;
import android.view.LayoutInflater;
import android.view.View;
import android.view.View.OnClickListener;
import android.view.ViewGroup;
```

```java
import android.widget.ImageButton;
import android.widget.Toast;
public class TitleFragment extends Fragment
{
    private ImageButton mLeftMenu;
    @Override
    public View onCreateView(LayoutInflater inflater, ViewGroup container,
            Bundle savedInstanceState)
    {
        View view=inflater.inflate(R.layout.titlefragment, container, false);
        mLeftMenu= (ImageButton) view.findViewById(R.id.id_title_left_btn);
        mLeftMenu.setOnClickListener(new OnClickListener()
        {
            @Override
            public void onClick(View v)
            {
                Toast.makeText(getActivity(),
                        "我是 TitleFragment",
                        Toast.LENGTH_SHORT).show();
            }
        });
        return view;
    }
}
```

(3) contentfragment.xml

```xml
<?xml version="1.0" encoding="utf-8"?>
<LinearLayout xmlns:android="http://schemas.android.com/apk/res/android"
    android:layout_width="match_parent"
    android:layout_height="match_parent"
    android:orientation="vertical">

    <TextView
        android:layout_width="fill_parent"
        android:layout_height="fill_parent"
        android:gravity="center"
        android:text="我是 ContentFragment"
        android:textSize="20sp"
        android:textStyle="bold" />

</LinearLayout>
```

(4) ContentFragment 类文件

```java
package com.example.android_demo5_8;
```

```java
import android.app.Fragment;
import android.os.Bundle;
import android.view.LayoutInflater;
import android.view.View;
import android.view.ViewGroup;

public class ContentFragment extends Fragment
{
    @Override
    public View onCreateView(LayoutInflater inflater, ViewGroup container,
            Bundle savedInstanceState)
    {
        return inflater.inflate(R.layout.contentfragment, container, false);
    }
}
```

(5) activity_main.xml

```xml
<RelativeLayout xmlns:android="http://schemas.android.com/apk/res/android"
    xmlns:tools="http://schemas.android.com/tools"
    android:layout_width="match_parent"
    android:layout_height="match_parent">
    <fragment
        android:id="@+id/id_fragment_title"
        android:name="com.example.android_demo5_8.TitleFragment"
        android:layout_width="fill_parent"
        android:layout_height="45dp" />
    <fragment
        android:layout_below="@id/id_fragment_title"
        android:id="@+id/id_fragment_content"
        android:name="com.example.android_demo5_8.ContentFragment"
        android:layout_width="fill_parent"
        android:layout_height="fill_parent" />
</RelativeLayout>
```

(6) MainActivity 类文件

```java
package com.example.android_demo5_8;
import android.app.Activity;
import android.os.Bundle;
import android.view.Window;
public class MainActivity extends Activity
{
    @Override
    protected void onCreate(Bundle savedInstanceState)
    {
```

```
        super.onCreate(savedInstanceState);
        requestWindowFeature(Window.FEATURE_NO_TITLE);
        setContentView(R.layout.activity_main);
    }
}
```

静态使用 Fragment 实例的运行效果如图 5-20 所示。

3. 动态使用 Fragment

为了动态使用 Fragment，修改上面实例中的 Activity 的布局文件，在其中使用了一个 FrameLayout，并添加了 4 个按钮；同时增加了一个 Fragment 类和它的布局文件。当用户单击 Activity 中底部的按钮时，会动态切换 Activity 界面中间的 Fragment。

MainActivity 类修改如下。

图 5-20 静态使用 Fragment 实例运行效果

```java
package com.example.android_demo5_8;

import android.app.Activity;
import android.app.FragmentManager;
import android.app.FragmentTransaction;
import android.os.Bundle;
import android.view.View;
import android.view.View.OnClickListener;
import android.view.Window;
import android.widget.ImageButton;

public class MainActivity extends Activity implements OnClickListener
{
    private ImageButton butContent;
    private ImageButton butFriend;
    private ContentFragment fContent;
    private FriendFragment fFriend;
    @Override
    protected void onCreate(Bundle savedInstanceState)
    {
        super.onCreate(savedInstanceState);
        requestWindowFeature(Window.FEATURE_NO_TITLE);
        setContentView(R.layout.activity_main);
        //初始化控件和声明事件
        butContent= (ImageButton) findViewById(R.id.but1);
        butFriend= (ImageButton) findViewById(R.id.but2);
```

```java
        butContent.setOnClickListener(this);
        butFriend.setOnClickListener(this);
        //设置默认的 Fragment
        setDefaultFragment();
    }
    private void setDefaultFragment()
    {
        FragmentManager fm=getFragmentManager();
        FragmentTransaction transaction=fm.beginTransaction();
        fContent=new ContentFragment();
        transaction.replace(R.id.id_content, fContent);
        transaction.commit();
    }
    @Override
    public void onClick(View v)
    {
        FragmentManager fm=getFragmentManager();
        //开启 Fragment 事务
        FragmentTransaction transaction=fm.beginTransaction();
        switch (v.getId())
        {
        case R.id.but1:
            if (fContent==null)
            {
                fContent=new ContentFragment();
            }
            //使用当前 Fragment 的布局替代 id_content 的控件
            transaction.replace(R.id.id_content, fContent);
            break;
        case R.id.but2:
            if (fFriend==null)
            {
                fFriend=new FriendFragment();
            }
            transaction.replace(R.id.id_content, fFriend);
            break;
        }
        //事务提交
        transaction.commit();
    }
}
```

可以看到我们使用 FragmentManager 对 Fragment 进行了动态的加载,这里使用的是 replace()方法。代码中间还有两个 Fragment 的子类,ContentFragment 上面已经见过,FriendFragment 和 ContentFragment 基本相同,就是显示的文字不一样。

MainActivity 的布局文件如下。

```xml
<RelativeLayout xmlns:android="http://schemas.android.com/apk/res/android"
    xmlns:tools="http://schemas.android.com/tools"
    android:layout_width="match_parent"
    android:layout_height="match_parent">

    <fragment
        android:id="@+id/id_fragmenttitle"
        android:name="com.example.android_demo5_8.TitleFragment"
        android:layout_width="fill_parent"
        android:layout_height="45dp" />
    <include
        android:id="@+id/id_bottombar"
        android:layout_width="fill_parent"
        android:layout_height="55dp"
        android:layout_alignParentBottom="true"
        layout="@layout/bottombar" />
    <FrameLayout
        android:id="@+id/id_content"
        android:layout_width="fill_parent"
        android:layout_height="fill_parent"
        android:layout_above="@id/id_bottombar"
        android:layout_below="@id/id_fragmenttitle" />
</RelativeLayout>
```

底部按钮布局文件 bottombar.xml 如下。

```xml
<?xml version="1.0" encoding="utf-8"?>
<LinearLayout xmlns:android="http://schemas.android.com/apk/res/android"
    android:layout_width="match_parent"
    android:layout_height="match_parent"
    android:layout_gravity="center_horizontal"
    android:background="@drawable/actionbarcolor"
    android:gravity="center_horizontal"
    android:orientation="horizontal">

    <ImageButton
        android:id="@+id/but1"
        android:layout_width="50dp"
        android:layout_height="50dp"
        android:layout_marginBottom="5dp"
        android:layout_marginLeft="25dp"
        android:layout_marginRight="25dp"
```

```xml
            android:layout_marginTop="5dp"
            android:background="@drawable/edit" />

        <ImageButton
            android:id="@+id/but2"
            android:layout_width="50dp"
            android:layout_height="50dp"
            android:layout_marginBottom="5dp"
            android:layout_marginLeft="25dp"
            android:layout_marginRight="25dp"
            android:layout_marginTop="5dp"
            android:background="@drawable/send"/>

        <ImageButton
            android:id="@+id/but3"
            android:layout_width="50dp"
            android:layout_height="50dp"
            android:layout_marginBottom="5dp"
            android:layout_marginLeft="25dp"
            android:layout_marginRight="25dp"
            android:layout_marginTop="5dp"
            android:background="@drawable/help"/>

        <ImageButton
            android:id="@+id/but4"
            android:layout_width="50dp"
            android:layout_height="50dp"
            android:layout_marginBottom="5dp"
            android:layout_marginLeft="25dp"
            android:layout_marginRight="25dp"
            android:layout_marginTop="5dp"
            android:background="@drawable/quit"/>

</LinearLayout>
```

动态使用 Fragment 实例的运行效果如图 5-21 所示。

5.8.3 Android 支持包

如果使用 Android 3.0 以下的版本，需要引入 android-support-v4.jar 支持包，然后让 Activity 继承 FragmentActivity，再通过 getSupportFragmentManager 获得 FragmentManager 进行下一步操作。

图 5-21 动态使用 Fragment 实例运行效果

习 题 5

1. 阅读下面的代码,分析其完成的功能,在【1】、【2】处填写上正确的代码。

```
public class MainActivity extends Activity {
    private ListView mListView;
    private List<String>list;
    @Override
    protected void onCreate(Bundle savedInstanceState) {
        super.onCreate(savedInstanceState);
        setContentView(R.layout.activity_main);
        mListView= (ListView) findViewById(R.id.listView1);
        list=new ArrayList<String>();
        final ArrayAdapter<String> _ArrayAdapter=【1】
        mListView.setAdapter(【2】);
        mListView.setOnItemClickListener(new OnItemClickListener()
        {
            @Override
            public void onItemClick(AdapterView<?>parent,
                    View view,int position, long id) {
                Toast.makeText(MainActivity.this,
                    parent.getItemAtPosition(position).toString(),
                    Toast.LENGTH_SHORT).show();
            }
        }
    }
}
```

2. 如下代码用来初始化进度条对话框,请在【1】、【2】、【3】、【4】、【5】号处填写上正确的代码。

```
public void initProgressDialog(){
    dailog=new ProgressDialog(MainActivity.this);
```

```
//设置水平样式进度条
dailog.setProgressStyle(【1】);
dailog.setTitle("进度条对话框");
dailog.setMessage("这是一个进度条对话框");
//设置对话框图标
    【2】
//设置按返回键不能退出对话框
    【3】
//设置进度为100
    【4】
dailog.setButton("确定",
    new android.content.DialogInterface.OnClickListener() {
        @Override
        public void onClick(DialogInterface arg0,int arg1) {
            dialog.cancal();
        }
    });
//显示对话框
    【5】
}
```

3. ArrayAdapter 和 SimpleAdapter 的作用是什么？有什么区别？
4. Android 的常用对话框有哪些？
5. 什么是 Fragment？说明其生命周期。

第 6 章

Android 多媒体

本章主要介绍了 Android 的图形绘制，多媒体播放和动画开发。通过本章的学习，读者可以了解如何通过自定义 View 进行 2D 图形绘制、播放音频和视频文件、使用 Android 提供的 Camera 类中的 API 实现拍照功能、设计逐帧动画和补间动画效果。

6.1 Android 的图形绘制

Android 图形绘制的基本方法是，先创建一个继承 View 或 SurfaceView 的类，然后重写其 onDraw()方法，在 onDraw()方法中会传递 Canvas 对象，该对象为画布对象，很多图形都是通过它的 drawXXX()方法进行绘制的，最后进行图形界面刷新。Android 用于图形绘制常用的类有 Canvas、Paint、Color 和 Path 等。

6.1.1 Canvas

Canvas 对象在 Java 和 JavaScript 中就已存在，Canvas 就像手机中的画布，可以在其上绘制图形或者图片。

Canvas 的常用方法如下。

1. Canvas()

创建一个空的画布，可以使用 setBitmap()方法来设置绘制具体的画布。

1) Canvas(Bitmap bitmap)

以 bitmap 对象创建一个画布，则将内容都绘制在 bitmap 上，因此 bitmap 不能为空（null）。

2) Canvas(GL gl)

在绘制 3D 效果时使用，与 OpenGL 相关。

2. translate(float dx, float dy)

移动原点。默认的原点在左上角，即(0,0)，可以通过 translate 方法移动原点。

3. rotate(float degree)

旋转指定角度的画布。

4. skew(float sx,float sy)

设置倾斜的值。

5. scale(float sx,float sy)

增减图形在 Canvas 中的像素数目,对形状、位图进行缩小或者放大。

6. drawColor(int color)

设置 Canvas 的背景颜色。

7. setBitmap()

设置具体画布。

8. clipRect()

设置显示区域,即设置裁剪区。

9. isOpaque()

检测是否支持透明。

10. void drawBitmap(Bitmap bitmap,Rect src,Rect dst,Paint paint)

贴图,参数一是 Bitmap 对象,参数二是源区域,参数三是目标区域,参数四是 Paint 画刷对象。

11. void drawLine(float startX,float startY,float stopX,float stopY,Paint paint)

画线,参数一为起始点的 x 轴位置,参数二为起始点的 y 轴位置,参数三为终点的 x 轴水平位置,参数四为 y 轴垂直位置,最后一个参数为 Paint 画刷对象。

12. void drawText(String text,float x,float y,Paint paint)

绘制文本,参数一是 String 类型的文本,参数二是起始点的 x 轴位置,参数三是起始点的 y 轴位置,参数四是 Paint 对象。

13. drawPosText(char[] text,int index,int count,float[] pos,Paint paint)

在指定的位置 pos 上绘制文字,从 text 数组中的 index 开始,共 count 个。

14. void drawPoint(float x,float y,Paint paint)

画点,参数一为点的 x 轴位置,参数二为点的 y 轴位置,第三个参数为 Paint 对象。

15. void drawRect(RectF rect,Paint paint)

绘制区域,参数一 RectF 为一个区域。

6.1.2 Paint

Paint 类拥有样式与颜色信息,主要是有关于如何绘制几何图形、文字及位图的方法。Paint 的常用方法如下。

1. void setARGB(int a,int r,int g,int b)

设置 Paint 对象颜色,参数一为 alpha 透明通道。

2. void setAlpha(int a)

设置 alpha 透明度,范围为 0~255。

3. void setAntiAlias(boolean aa)

是否抗锯齿。

4. void setColor(int color)

设置颜色,Android 内部定义的有 Color 类包含了一些常见颜色定义。

5. void setFakeBoldText(boolean fakeBoldText)

是否设置伪粗体文本。

6. void setLinearText(boolean linearText)

设置线性文本。

7. PathEffect setPathEffect(PathEffect effect)

设置路径效果。

8. Rasterizer setRasterizer(Rasterizer rasterizer)

设置光栅化。

9. void setTextAlign(Paint.Align align)

设置文本对齐。

10. void setTextScaleX(float scaleX)

设置文本缩放倍数,1.0f 为原始大小。

11. void setTextSize(float textSize)

设置字体大小。

12. Typeface setTypeface(Typeface typeface)

设置字体效果,Typeface 包含了字体的类型、粗细,还有倾斜、颜色等。

13. void setUnderlineText(boolean underlineText)

设置下划线。

14. setDither(boolean dither)

设定是否使用图像抖动处理,使绘制出来的图片颜色更加平滑和饱满,图像更加清晰。

15. setFilterBitmap(boolean filter)

如果该项设置为 true,则图像在动画进行中会过滤掉对 Bitmap 图像的优化操作,加快显示速度,本设置项依赖于 dither 和 xfermode 的设置。

16. setStyle(Paint.Style style)

设置画笔的样式,其参数值可以为 FILL、FILL_OR_STROKE 或 STROKE。

17. setStrokeCap(Paint.Cap cap)

当画笔样式为 STROKE 或 FILL_OR_STROKE 时,设置笔刷的图形样式,如圆形样式 Cap.ROUND 或方形样式 Cap.SQUARE。

18. setSrokeJoin(Paint.Join join)

设置绘制时各图形的结合方式,如平滑效果等。

19. setStrokeWidth(float width)

当画笔样式为 STROKE 或 FILL_OR_STROKE 时,设置笔刷的粗细度。

20. setXfermode(Xfermode xfermode)

设置图形重叠时的处理方式,如合并、取交集或并集,经常用来制作橡皮的擦除效果。

21. setMaskFilter(MaskFilter maskfilter)

设置 MaskFilter,可以用不同的 MaskFilter 实现滤镜效果,如滤化、立体、浮雕等效果。

22. setColorFilter（ColorFilter colorfilter）

设置颜色过滤器，可以在绘制颜色时实现不同颜色的变换效果。

23. setPathEffect（PathEffect effect）

设置绘制路径的效果，如点画线等。

24. setShader（Shader shader）

设置图像效果，使用 Shader 可以绘制出各种渐变效果。

25. setShadowLayer（float radius，float dx，float dy，int color）

在图形下面设置阴影层，产生阴影效果，radius 为阴影的角度，dx 和 dy 为阴影在 x 轴和 y 轴上的距离，color 为阴影的颜色。

6.1.3　温度计绘图案例

下面将通过一个案例演示说明，使用 Paint 和 Canvas 绘制图形的方法。案例实现的功能是，接收用户输入的温度值，在 Activity 界面中绘制出一个温度计形状，模拟显示出温度所在的可读值。代码中使用了 SurfaceView 控件。案例代码如下。

1. Activity 布局文件 activity_main.xml

```
<LinearLayout xmlns:android="http://schemas.android.com/apk/res/android"
    xmlns:tools="http://schemas.android.com/tools"
    android:layout_width="match_parent"
    android:layout_height="match_parent"
    android:gravity="center_horizontal"
    android:orientation="vertical"
    tools:context="com.example.android_demo6_1.MainActivity">

    <LinearLayout
        android:id="@+id/linearLayout1"
        android:layout_width="match_parent"
        android:layout_height="wrap_content"
        android:layout_margin="5dp"
        android:layout_marginTop="30dp"
        android:orientation="horizontal">

        <TextView
            android:id="@+id/txt_temp"
            android:layout_width="wrap_content"
            android:layout_height="wrap_content"
```

```xml
            android:text="温度" />

        <EditText
            android:id="@+id/tv_temperature_number"
            android:layout_width="wrap_content"
            android:layout_height="wrap_content"
            android:ems="10">
            <requestFocus />
        </EditText>

        <Button
            android:id="@+id/but"
            android:layout_width="50dp"
            android:layout_height="30dp"
            android:onClick="but_Onclick"
            android:text="显示"
            android:textSize="12sp" />

    </LinearLayout>

    <SurfaceView
        android:id="@+id/sv_temperature"
        android:layout_width="match_parent"
        android:layout_height="match_parent"
        android:layout_margin="5dp" />

</LinearLayout>
```

2. MainActivity 类文件

```java
package com.example.android_demo6_1;

import android.app.Activity;
import android.graphics.Canvas;
import android.graphics.Color;
import android.graphics.Paint;
import android.view.SurfaceHolder.Callback;
import android.os.Bundle;
import android.util.Log;
import android.view.Menu;
import android.view.MenuItem;
import android.view.SurfaceHolder;
import android.view.SurfaceView;
import android.view.View;
```

```java
import android.widget.EditText;

public class MainActivity extends Activity implements Callback {

    private EditText temp;
    private SurfaceView mSurface;
    private SurfaceHolder mHolder;
    private Paint mPaint, paintCircle, paintLine;

    @Override
    protected void onCreate(Bundle savedInstanceState) {
        super.onCreate(savedInstanceState);
        setContentView(R.layout.activity_main);
        temp= (EditText)findViewById(R.id.tv_temperature_number);
        mSurface= (SurfaceView)findViewById(R.id.sv_temperature);
        mHolder=mSurface.getHolder();
        mHolder.addCallback(this);
    }

    public void but_Onclick(View v){
        try{
            drawtemp(temp.getText().toString());
        }catch(Exception e){
            Log.e("temp",e.getMessage());
        }
    }

    //画出温度计并显示温度
    private void drawtemp(String temp){
        //声明画笔
        mPaint=new Paint();
        //声明圆形画笔
        paintCircle=new Paint();
        //声明直线画笔
        paintLine=new Paint();
        //获取 canvas
        Canvas canvas=mHolder.lockCanvas();
        //获得 Y 轴数据
        int y=600 - (Integer.valueOf(temp) * 4);
        //设置画笔颜色
        mPaint.setColor(Color.WHITE);
        //绘制矩形
        canvas.drawRect(300, 200, 350, 600, mPaint);
        //设置圆形画笔颜色
```

```java
        paintCircle.setColor(Color.RED);
        //绘制圆形
        canvas.drawCircle(325, 645, 50, paintCircle);
        //绘制矩形
        canvas.drawRect(300, y, 350, 600, paintCircle);
        //设置直线画笔颜色
        paintLine.setColor(Color.BLUE);
        //设置Y轴值
        int lineY=600;
        //设置数字变量
        int num=0;
        //判断Y轴直线是否超过方位
        while (lineY>=200) {
            //绘制温度计直线
            canvas.drawLine(350, lineY, 355, lineY, mPaint);
            //每隔40绘制一条直线
            if (lineY%40==0 ) {
                //设置直线
                canvas.drawLine(350, lineY, 360, lineY, paintLine);
                //绘制文本
                canvas.drawText(String.valueOf(num), 362, lineY+4, mPaint);
                num+=10;
            }
            lineY -=4;
        }
        mHolder.unlockCanvasAndPost(canvas);            //更新屏幕显示内容
    }


    @Override
    public boolean onCreateOptionsMenu(Menu menu) {
        //Inflate the menu; this adds items to the action bar if it is present.
        getMenuInflater().inflate(R.menu.main, menu);
        return true;
    }


    @Override
    public boolean onOptionsItemSelected(MenuItem item) {
        //Handle action bar item clicks here. The action bar will
        //automatically handle clicks on the Home/Up button, so long
        //as you specify a parent activity in AndroidManifest.xml.
        int id=item.getItemId();
        if (id==R.id.action_settings) {
            return true;
```

```
        }
        return super.onOptionsItemSelected(item);
    }

    @Override
    public void surfaceCreated(SurfaceHolder holder) {
        //TODO Auto-generated method stub

    }

    @Override
    public void surfaceChanged(SurfaceHolder holder, int format, int width,
            int height) {
        //TODO Auto-generated method stub

    }

    @Override
    public void surfaceDestroyed(SurfaceHolder holder) {
        //TODO Auto-generated method stub

    }
}
```

3. 案例运行效果

案例运行效果如图 6-1 所示。

6.1.4 Bitmap

Bitmap 是 Android 图像处理中最重要的类之一。利用 Bitmap 可以获取图像文件信息，进行图像剪切、旋转、缩放等操作，并可以以指定格式保存图像文件。

1. 创建和获取位图

可以使用下面几种方法获取位图：
1）通过 BitmapFactory 的各种静态方法
（1）根据资源文件创建：

```
Bitmap bmp=BitmapFactory.decodeResource(this.getResources(), R.drawable.bmsrc);
```

图 6-1 温度计绘图案例运行效果

（2）根据图片创建：

```
Bitmap bmp=BitmapFactory.decodeFile("/sdcard/dcoim/pets.jpeg");
```

2) 通过 Drawable 对象

使用 BitmapDrawable(InputStream is)构造一个 BitmapDrawable；

使用 BitmapDrawable 类的 getBitmap()获取得到位图。

3) 通过资源

需要先得到 Resources 对象，然后调用 openRawResource()方法获取输入流，并将输入流传给一个 BitmapDrawable 对象，然后调用 Bitmap 对象的 getBitmap()方法得到位图。例如：

```
//得到 Resources 对象
Resources r=getApplicationContext().getResources();
//以数据流的方式读取资源
InputStream is=r.openRawResource(R.drawable.my_background_image);
BitmapDrawable bmpDraw=new BitmapDrawable(is);
Bitmap bmp=bmpDraw.getBitmap();
```

2. 获取位图信息

通过 Bitmap 的方法可以获取位图信息，例如位图大小、是否包含透明度、颜色格式等。有以下两点需要注意。

(1) 在 Bitmap 中对 RGB 颜色格式使用 Bitmap.Config 定义，仅包括 ALPHA_8、ARGB_4444、ARGB_8888 和 RGB_565 四种，颜色格式说明如下。

① ALPHA_8：图形参数由一个字节来表示，是一种 8 位的位图。

② ARGB_4444：图形的参数应该由两个字节来表示，是一种 16 位的位图。

③ ARGB_8888：图形的参数应该由 4 个字节来表示，是一种 32 位的位图。

④ RGB_565：图形的参数应该由两个字节来表示，是一种 16 位的位图。

(2) Bitmap 还提供了 compress()接口来压缩图片，不过 Android SDK 只支持 PNG 和 JPG 格式的压缩。

3. 显示位图

显示位图需要使用核心类 Canvas，可以直接通过 Canvas 类的 drawBirmap()显示位图，或者借助于 BitmapDrawable 将 Bitmap 绘制到 Canvas。

4. 缩放和旋转位图

Android 提供了以下两种位图缩放的方法。

(1) drawBitmap(Bitmap bitmap,Rect src,Rect dst,Paint paint)

将一个位图按照需求重画一遍。

(2) createBitmap(Bitmap source, int x, int y, int width, int height, Matrix m, boolean filter)

在原有位图的基础上，缩放原位图，创建一个新的位图。

6.1.5 Matrix

通过 Matrix,可以非常容易地控制 Android 绘图坐标的位移、旋转、缩放等功能。Matrix 的操作,共分为 translate(平移)、rotate(旋转)、scale(缩放)和 skew(倾斜)4 种。每一种变换在 Android 的 API 里都提供了 set、post 和 pre 三种操作方式,除了 translate,其他三种操作都可以指定中心点。

(1) set 是直接设置 Matrix 的值,每 set 一次,整个 Matrix 数组都会改变。

(2) post 是后乘,即当前的矩阵乘以参数给出的矩阵。可以连续多次使用 post,来完成所需的整个变换。

(3) pre 是前乘,参数给出的矩阵乘以当前的矩阵。所以操作是在当前矩阵的最前面发生的,如:

M.preConcat(other)>M'=M*other, M.postConcat(other)>M'=other*M

旋转、缩放和倾斜都可以围绕一个中心点来进行,如果不指定,默认情况下围绕(0,0)点来进行。其他还有像 setRotate、setSinCos、setScale、setTranslate 等方法可以应用。

6.1.6 图片缩放功能案例

一个使用 Matrix 的案例,使用 Matrix 实现图片的放大和缩小功能。

1. MainActivity 类如下

```java
package com.example.android_demo6_2;

import android.app.Activity;
import android.graphics.Bitmap;
import android.graphics.BitmapFactory;
import android.graphics.Matrix;
import android.os.Bundle;
import android.view.Menu;
import android.view.MenuItem;
import android.view.View;
import android.widget.Button;
import android.widget.ImageView;

public class MainActivity extends Activity {
    Matrix matrix=new Matrix();
    Bitmap bm;

    @Override
    public void onCreate(Bundle savedInstanceState) {
```

```java
        super.onCreate(savedInstanceState);
        setContentView(R.layout.activity_main);
        final ImageView iv=(ImageView) findViewById(R.id.imageView1);
        Button btn1=(Button) findViewById(R.id.button1);
        bm=BitmapFactory.decodeResource(getResources(),
            R.drawable.ic_launcher);
        btn1.setOnClickListener(new View.OnClickListener() {
            @Override
            public void onClick(View v) {
                matrix.postScale(1.1f, 1.1f);
                Bitmap bitmap=Bitmap.createBitmap(bm, 0, 0, bm.getWidth(),
                    bm.getHeight(), matrix, true);
                iv.setImageBitmap(bitmap);
            }
        });

        Button btn2=(Button) findViewById(R.id.button2);
        btn2.setOnClickListener(new View.OnClickListener() {
            @Override
            public void onClick(View v) {
                matrix.postScale(0.9f, 0.9f);
                Bitmap bitmap=Bitmap.createBitmap(bm, 0, 0, bm.getWidth(),
                    bm.getHeight(), matrix, true);
                iv.setImageBitmap(bitmap);
            }
        });
    }

    @Override
    public boolean onCreateOptionsMenu(Menu menu) {
        //Inflate the menu; this adds items to the action bar if it is present.
        getMenuInflater().inflate(R.menu.main, menu);
        return true;
    }

    @Override
    public boolean onOptionsItemSelected(MenuItem item) {
        int id=item.getItemId();
        if (id==R.id.action_settings) {
            return true;
        }
        return super.onOptionsItemSelected(item);
    }
}
```

2. Activity 的界面布局文件 activity_main.xml

```xml
<?xml version="1.0" encoding="utf-8"?>
<LinearLayout xmlns:android="http://schemas.android.com/apk/res/android"
    android:layout_width="fill_parent"
    android:layout_height="fill_parent"
    android:layout_gravity="center"
    android:orientation="vertical">

    <LinearLayout
        android:id="@+id/linearLayout1"
        android:layout_width="wrap_content"
        android:layout_height="wrap_content"
        android:layout_gravity="center"
        android:orientation="horizontal">

        <Button
            android:id="@+id/button1"
            android:layout_width="wrap_content"
            android:layout_height="wrap_content"
            android:text="放大">
        </Button>

        <Button
            android:id="@+id/button2"
            android:layout_width="wrap_content"
            android:layout_height="wrap_content"
            android:text="缩小">
        </Button>
    </LinearLayout>

    <ImageView
        android:id="@+id/imageView1"
        android:layout_width="wrap_content"
        android:layout_height="wrap_content"
        android:layout_gravity="center"
        android:src="@drawable/ic_launcher">
    </ImageView>

</LinearLayout>
```

3. 案例运行效果

案例的运行效果如图 6-2 所示,当单击"放大"按

图 6-2 图片缩放功能运行效果

钮时，图片会放大，反之会缩小。

6.2 Android 多媒体基础

Android 的多媒体框架包含了对各种通用的媒体类型的支持，例如 MPEG4、H.264、MP3、AAC、AMR、JPG、PNG、GIF 等，通过 Android 的多媒体框架，开发者可以轻易地在应用中集成音频和视频处理功能，播放来源于网络的多媒体流、本地文件、应用程序资源以及获取到的各种音频、视频数据。

Android 提供了 MediaPlayer 类来实现多媒体播放功能，而 MediaPlayer 类在底层上是基于 OpenCore(也叫 PacketVideo)库实现的。本章的多媒体部分就是在讲解，如何通过 MediaPlayer 类来实现 Android 设备的多媒体播放功能。

需要注意，Android 中只能通过标准的输出设备播放音频，如扬声器或蓝牙耳机，而这些设备被设定为是通话功能独占的，所以打电话时不能播放音频文件。

6.2.1 基本类

Android 多媒体框架中用来播放音频和视频的两个类如下。
（1）MediaPlayer：用来实现音频和视频播放功能，提供播放所需要的所有基础 API。
（2）AudioManager：用来管理音频资源和音频输出设备。

6.2.2 权限声明

在 Android 的多媒体开发中，有很多情况需要在 AndroidManifest.xml 中添加相应的权限声明。

1. 网络权限

如果需要播放来自网络的多媒体流，需要添加网络访问权限声明：

`<uses-permission android:name="android.permission.INTERNET" />`

2. 访问 SDcard 卡权限

如果需要播放存储在 SDcard 卡上的多媒体文件，则需要添加权限：

`<uses-permission android:name="android.permission.READ_EXTERNAL_STORAGE"/>`

3. 唤醒锁权限

在播放多媒体文件时，如果需要让屏幕变暗或处理器睡眠，或者说需要调用 MediaPlayer 的 setScreenOnWhilePlaying()或 MediaPlayer.setWakeMode()方法，则必须加入如下权限：

```
<uses-permission android:name="android.permission.WAKE_LOCK"/>
```

6.2.3 Android 多媒体核心 OpenCore

OpenCore 是 Android 的多媒体核心，它位于 Android 体系结构中的本地库层，它是一个使用 C++ 实现的、代码量庞大且功能齐全、性能优良、可供其他开发者在其上做扩展开发的多媒体库。Android 中的 MediaPlayer 类就是基于该库实现的。实际上，开发者调用 MediaPlayer 类中的方法实现多媒体播放功能时，就是间接地调用了该库的一系列多媒体播放函数。

OpenCore 定义了全功能的操作系统移植层，各种基本的功能均被封装成类的形式，各层次之间的接口很多使用继承等方式。OpenCore 是一个多媒体框架，从宏观上看，它主要包含了以下两大方面的内容。

（1）PVPlayer：提供媒体播放器的功能，完成各种音频（Audio）、视频（Video）流的回放（Playback）功能。

（2）PVAuthor：提供媒体流记录功能，完成各种音频（Audio）、视频（Video）流以及静态图像捕获功能。

PVPlayer 和 PVAuthor 以 SDK 的形式提供给开发者，可以在这个 SDK 之上构建多种应用程序和服务（在移动终端中常使用的多媒体应用程序，例如媒体播放器、照相机、录像机、录音机等）。

为了更好地组织整体架构，OpenCore 软件层次在宏观上分成了以下几个层次。

（1）OSCL（Operating System Compatibility Library，操作系统兼容库）：为了更好地在不同操作系统移植，包含了一些操作系统底层的操作，如基本数据类型、配置、字符串工具、IO、错误处理、线程等内容，类似一个基础的 C++ 库。

（2）PVMF（PacketVideo Multimedia Framework，PV 多媒体框架）：在框架内实现一个文件解析（Parser）和组成（Composer）、编解码 NODE，也可以继承其通用接口，在用户层实现一些 NODE。

（3）PVPlayer Engine：PVPlayer 引擎。

（4）PVAuthor Engine：PVAuthor 引擎。

事实上，OpenCore 中包含的内容非常多。从播放的角度看，PVPlayer 的输入是文件或者网络媒体流，输出是音频和视频的输出设备，其基本功能包含了媒体流控制、文件解析、音频视频流的解码等方面的内容。除了从文件中播放媒体文件之外，还包含了与网络相关的 RTSP 流（Real Time Stream Protocol，实时流协议）。在媒体流记录方面，PVAuthor 的输入是照相机、麦克风等设备，输出是各种文件，包含了流的同步、音频视频流的编码以及文件的写入等功能。

在使用 OpenCore SDK 的时候，有可能需要在应用程序层实现一个适配器，然后在适配器之上实现具体的功能，对于 PVMF 的 NODE 也可以基于通用接口，在上层实现，并以插件的形式使用。

6.2.4 MediaPlayer 类

1. MediaPlayer 对象的生命周期

MediaPlayer 对象对音频和视频的播放过程并不是单一的播放或停止,而是被实现成由多个状态组成的一整套机制,这套机制就是 MediaPlayer 的生命周期,图 6-3 就是 MediaPlayer 的生命周期示意图。其中的椭圆代表生命周期中的各个状态,箭头代表各个方法,单箭头的是同步方法,双箭头的是异步方法。

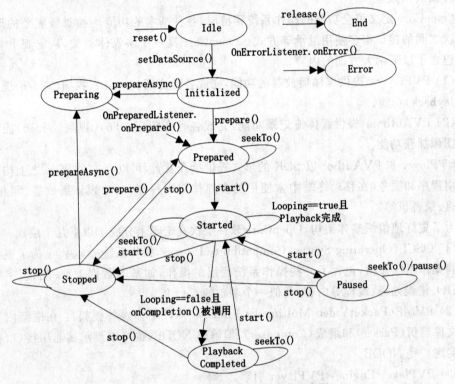

图 6-3 MediaPlayer 的生命周期示意图

如图 6-3 所示,当一个 MediaPlayer 对象被新建或者调用 reset()方法之后,它处于空闲(Idle)状态,在调用 release()方法后,才会处于结束(End)状态。一些常用的播放控制操作,可能因为音频、视频格式不被支持或者质量较差以及流超时等而导致错误,这时可以通过注册 setOnErrorListener(android.media.MediaPlayer.OnErrorListener)方法实现 OnErrorListener.onError()方法来监控这些错误。当发生错误后,可以使用 reset()方法来恢复错误。

任何 MeidaPlayer 对象都必须处于准备(Prepared)状态,然后才开始播放。要开始播放 MeidaPlayer 对象都必须成功调用 start()方法,可以通过 isPlaying()方法来检测当前是否正在播放,还可以通过 setLooping(boolean)方法来设置是否循环播放。

当 MeidaPlayer 对象在播放时,可以进行暂停和停止等操作,pause()方法暂停播放,

stop()方法停止播放。处于暂停(Paused)状态可以通过 start()方法来恢复播放,但是处于停止(Stopped)状态则必须先调用 prepare()方法使之处于准备状态,然后通过 start()方法来开始播放。

2. MediaPlayer 的常用方法

MediaPlayer 类中常用的方法如表 6-1 所示。

表 6-1 MediaPlayer 的常用方法

MediaPlayer 的常用方法	方 法 说 明
static MediaPlayer create(Context context,Uri uri)	通过 URI 创建一个多媒体播放器
static MediaPlayer create(Context context, int resid)	通过资源 ID 创建一个多媒体播放器
static MediaPlayer create (Context context,Uri uri, SurfaceHolder holder)	通过 URI 和指定的 SurfaceHolder 创建一个多媒体播放器
int getCurrentPosition()	得到当前播放位置
int getDuration()	得到所播放文件的持续时间
int getVideoHeight()	得到视频的高度
int getVideoWidth()	得到视频的宽度
boolean isLooping()	是否循环播放
boolean isPlaying()	是否正在播放
void pause()	暂停
void prepare()	同步地准备播放或回放
void prepareAsync()	异步地准备播放或回放
void release()	释放 MediaPlayer 对象
void reset()	重置 MediaPlayer 对象
void seekTo(int msec)	指定播放的位置(以 ms 为单位的时间)
void setAudioStreamType(int streamtype)	指定流媒体的类型
void setDataSource(String path)	根据路径设置多媒体数据来源
void setDataSource(FileDescriptor fd)	根据 FileDescriptor 设置多媒体数据来源
void setDataSource (FileDescriptor fd, long offset, long length)	根据 FileDescriptor 设置多媒体数据来源,并指定偏移量和总量
void setDataSource(Context context,Uri uri)	根据 URI 设置多媒体数据来源
void setDisplay(SurfaceHolder sh)	设置用 SurfaceHolder 来显示多媒体
void setLooping(boolean looping)	设置是否循环播放
void setOnBufferingUpdateListener (MediaPlayer.OnBufferingUpdateListener listener)	监听事件,网络流媒体的缓冲监听
void setOnCompletionListener (MediaPlayer.OnCompletionListener listener)	监听事件,网络流媒体播放结束监听

续表

MediaPlayer 的常用方法	方 法 说 明
void setOnErrorListener (MediaPlayer.OnErrorListener listener)	监听事件，设置错误信息监听
void setOnVideoSizeChangedListener (MediaPlayer.OnVideoSizeChangedListener listener)	监听事件，视频尺寸监听
void setScreenOnWhilePlaying(boolean screenOn)	设置是否使用 SurfaceHolder 显示
void setVolume(float leftVolume, float rightVolume)	设置音量
void start()	开始播放
void stop()	停止播放

6.3 音 频 播 放

MediaPlayer 对象可以通过简单的设置来获取、解码、播放音频和视频资源，可以支持以下三种不同的多媒体来源：

（1）本地资源。

（2）内部的 URI 指向的资源，例如通过一个 ContentResolver（数据共享）对象获得的。

（3）外部的 URL 指向的资源，往往是从网络上获取到的多媒体流。

6.3.1 播放本地资源

可以像使用图片资源那样，把一些比较小的多媒体资源放在 Android 项目的 res/raw/文件夹中，这样就可以使用 R.raw.xxx 来找到这个多媒体资源文件。但是由于放在该路径下的文件会打包进 apk 安装包文件中，所以不要把太大的多媒体文件放在其中，例如几十 MB 或更大的视频文件。对于较大的音频或视频文件，可以将其放在 SD 卡中，或者直接放在服务器上。

下面的代码片段就可以播放位于 Android 项目的 res/raw/目录中的音频文件 sound_file。

```
MediaPlayer mediaPlayer=MediaPlayer.create(context, R.raw.sound_file);
//这种情况下,不用调用 prepare(),因为已经在 create()中已经完成
mediaPlayer.start();
```

6.3.2 播放内部资源

对于程序内部通过 URI 指定的多媒体资源，例如说从 ContentResolver 得到的资源，如下代码片段就可以播放：

```
Uri myUri=Uri.parse("content://..."); //初始化资源的 URI
MediaPlayer mediaPlayer=new MediaPlayer();
mediaPlayer.setAudioStreamType(AudioManager.STREAM_MUSIC);
try {
        mediaPlayer.setDataSource(
        getApplicationContext(), myUri);
        mediaPlayer.prepare();
        mediaPlayer.start();
} catch (IllegalArgumentException e) {
        e.printStackTrace();
} catch (SecurityException e) {
        e.printStackTrace();
} catch (IllegalStateException e) {
        e.printStackTrace();
} catch (IOException e) {
        e.printStackTrace();
}
```

6.3.3 播放网络资源

可以用如下的代码片段播放从网络上获取到的多媒体流。

```
String url="http://……";                //指定资源在网络上的地址
MediaPlayer mediaPlayer=new MediaPlayer();
mediaPlayer.setAudioStreamType(AudioManager.STREAM_MUSIC);
try {
        mediaPlayer.setDataSource(url);
        mediaPlayer.prepare();                //可能需要较长时间,因为需要缓冲
} catch (IllegalArgumentException e) {
        e.printStackTrace();
} catch (SecurityException e) {
        e.printStackTrace();
} catch (IllegalStateException e) {
        e.printStackTrace();
} catch (IOException e) {
        e.printStackTrace();
}
mediaPlayer.start();
```

以上代码播放的网络资源必须是可以下载的,也就是需要已经拥有了下载权限,才可以将其播放出来。

需要注意,使用 setDataSource()方法时,必须处理 IllegalArgumentException 和 IOException 异常,因为所使用的文件可能不存在。

6.4 简单音乐播放器案例

6.4.1 案例功能描述

案例将提供一个简单的音乐播放器,能播放本地资源的音频文件,只能实现播放、暂停和停止三个功能。

6.4.2 案例程序结构

案例中包括一个布局文件(activity_main.xml);一个 Activity 组件类(MainActivity 类),用于实现用户界面交互功能;一个 Service 组件类(MediaPlayoerService 类),用于控制媒体播放器。

6.4.3 案例的实现步骤和思路

(1) 创建 Android 项目。

(2) 创建 MainActivity 的布局文件 activity_main.xml。

布局中最外层容器为 RelativeLayout 布局,宽和高都是充满父容器,内部有三个按钮控件。

(3) 创建继承于 Service 服务类的 MediaPlayoerService 服务。

① 编写内部类 MyBind 继承 Binder,用于返回连接对象,内部编写方法控制播放器的播放、暂停、停止;

② 复写 onBind 方法,返回一个连接对象 MyBind;

③ 复写 onCreate 方法进行数据初始化;

④ 复写 onStartCommand 方法;

⑤ 复写 onUnbind 方法;

⑥ 复写 onDestroy 方法,在方法中释放媒体播放器。

(4) 编写 MainActivity 类文件。

① 在 onCreate 方法中,先调用父类的 onCreate 方法;

② 使用 setContentView 方法加载布局文件 activity_main.xml;

③ 获取界面的三个按钮控件,给每个按钮设置监听器;

④ 创建连接对象 ServiceConnection,复写 onServiceDisconnected 和 onServiceConnected 方法,在 onServiceConnected 方法中获取连接返回的 MyBind 对象;

⑤ 使用 bindService 方法连接服务;

⑥ 在按钮监听器中,根据不同的按钮调用 MyBind 的公开方法,来控制媒体播放器。

6.4.4 案例参考代码

(1) 布局文件 activity_main.xml 代码如下。

```xml
<?xml version="1.0" encoding="utf-8"?>
<!--主界面布局,最外面是相对布局 -->
<RelativeLayout xmlns:android="http://schemas.android.com/apk/res/android"
    xmlns:tools="http://schemas.android.com/tools"
    android:layout_width="match_parent"
    android:layout_height="match_parent"
    android:paddingBottom="@dimen/activity_vertical_margin"
    android:paddingLeft="@dimen/activity_horizontal_margin"
    android:paddingRight="@dimen/activity_horizontal_margin"
    android:paddingTop="@dimen/activity_vertical_margin"
    tools:context="pjy.caiwei.MainActivity">

    <!--布局中第一个控件 Button -->
    <Button
        android:id="@+id/chapter6_1_2_play"
        android:layout_width="match_parent"
        android:layout_height="wrap_content"
        android:layout_alignParentLeft="true"
        android:layout_alignParentTop="true"
        android:layout_marginTop="176dp"
        android:text="play" />

    <!--布局中第二个控件 Button -->
    <Button
        android:id="@+id/chapter6_1_2_pause"
        android:layout_width="match_parent"
        android:layout_height="wrap_content"
        android:layout_alignLeft="@+id/chapter6_1_2_play"
        android:layout_below="@+id/chapter6_1_2_play"
        android:text="pause" />

    <!--布局中第三个控件 Button -->
    <Button
        android:id="@+id/chapter6_1_2_stop"
        android:layout_width="match_parent"
        android:layout_height="wrap_content"
        android:layout_alignLeft="@+id/chapter6_1_2_pause"
        android:layout_below="@+id/chapter6_1_2_pause"
        android:text="stop" />
```

```
</RelativeLayout>
```

(2) Service 类 MediaPlayerService 代码如下。

```java
package com.example.android_demo6_3;

import android.app.Service;
import android.content.Intent;
import android.media.MediaPlayer;
import android.os.Binder;
import android.os.IBinder;
import android.util.Log;
/**
 * 服务类,控制媒体播放器
 */
public class MediaPlayerService extends Service {
    //媒体播放器
    private MediaPlayer mPlayer;
    //服务连接对象
    private MyBind mBind;
    /**
     * 内部类,服务绑定对象
     */
    public class MyBind extends Binder {
        //公开方法,控制 MediaPlayer 播放
        public void play() {
            if (mPlayer !=null) {
                if (!mPlayer.isPlaying()) {
                    mPlayer.start();
                }
            }
        }
        //公开方法,控制 MediaPlayer 暂停
        public void pause() {
            if (mPlayer !=null) {
                if (mPlayer.isPlaying()) {
                    mPlayer.pause();
                }
            }
        }
        //公开方法,控制 MediaPlayer 停止
        public void stop() {
            if (mPlayer !=null) {
                mPlayer.stop();
                try {
```

```java
                mPlayer=MediaPlayer.create(MediaPlayerService.this,
                    R.raw.audio);
                mPlayer.prepare();
            } catch (Exception e) {
                e.printStackTrace();
            }
        }
    }
    /**
     * 服务生命周期方法,服务绑定时调用
     */
    @Override
    public IBinder onBind(Intent intent) {
        Log.i("MyLog", "服务被绑定-------");
        return mBind;
    }
    /**
     * 服务生命周期方法,服务创建时调用
     */
    @Override
    public void onCreate() {
        super.onCreate();
        mPlayer=MediaPlayer.create(this, R.raw.audio);
        mBind=new MyBind();
        try {
            mPlayer.prepare();
        } catch (Exception e) {
            e.printStackTrace();
        }
    }
    /**
     * 服务生命周期方法,服务开启时调用
     */
    @Override
    public int onStartCommand(Intent intent, int flags, int startId) {
        return super.onStartCommand(intent, flags, startId);
    }
    /**
     * 服务生命周期方法,服务解除绑定时调用
     */
    @Override
    public boolean onUnbind(Intent intent) {
        return super.onUnbind(intent);
```

```java
    }
    /**
     *服务生命周期方法,服务销毁时调用
     */
    @Override
    public void onDestroy() {
        super.onDestroy();
        mPlayer.stop();
        mPlayer.release();
    }
}
```

(3) Activity 组件 MainActivity 类代码如下。

```java
package com.example.android_demo6_3;

import pjy.caiwei.MediaPlayerService.MyBind;
import android.app.Activity;
import android.content.ComponentName;
import android.content.Context;
import android.content.Intent;
import android.content.ServiceConnection;
import android.os.Bundle;
import android.os.IBinder;
import android.util.Log;
import android.view.View;
import android.view.View.OnClickListener;
import android.widget.Button;

/**
 *主 Activity 为控制类,用于控制界面的显示
 */
public class MainActivity extends Activity {
    //存储绑定状态
    boolean isbind=false;
    //Bind 对象
    MyBind mBind;
    //连接器
    ServiceConnection conn=new ServiceConnection() {
        public void onServiceDisconnected(ComponentName name) {
        }
        public void onServiceConnected(ComponentName name, IBinder service) {
            mBind= (MyBind) service;
            Log.i("MyLog", "服务被绑定");
```

```java
        }
    };
    /**
     * Activity组件的生命周期方法,在组件创建时调用,一般用于初始化信息
     */
    @Override
    protected void onCreate(Bundle savedInstanceState) {
        //调用父类方法,完成系统工作
        super.onCreate(savedInstanceState);
        //加载界面布局文件
        setContentView(R.layout.activity_main);
        //获取界面控件
        Button play= (Button) findViewById(R.id.chapter6_1_2_play);
        Button pause= (Button) findViewById(R.id.chapter6_1_2_pause);
        Button stop= (Button) findViewById(R.id.chapter6_1_2_stop);
         //设置按钮监听器
        MyOnClickListener listener=new MyOnClickListener();
        play.setOnClickListener(listener);
        pause.setOnClickListener(listener);
        stop.setOnClickListener(listener);
        Intent intent=new Intent();
        intent.setClass(MainActivity.this,MediaPlayerService.class);
        //绑定服务
        bindService(intent, conn, Context.BIND_AUTO_CREATE);
    }
    /**
     * Activity组件的生命周期方法,在销毁时调用,解除服务绑定
     */
    @Override
    protected void onDestroy() {
        super.onDestroy();
        unbindService(conn);
    }
    /**
     * 内部类,处理按钮单击事件
     */
    class MyOnClickListener implements OnClickListener {
        public void onClick(View v) {
            switch (v.getId()) {
                case R.id.chapter6_1_2_play:
                    mBind.play();
                    break;
                case R.id.chapter6_1_2_pause:
```

```
                    mBind.pause();
                    break;
                case R.id.chapter6_1_2_stop:
                    mBind.stop();
                    break;
            }
        }
    }
}
```

6.4.5 案例运行效果

简单音乐播放器运行效果如图 6-4 所示。

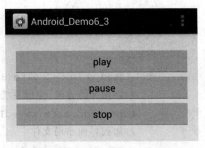

图 6-4 简单音乐播放器运行效果

6.5 视 频 播 放

Android 的 MediaPlayer 类主要用于播放音频，它没有提供输出图像的界面。因此需要使用 SurfaceView 控件，与 MediaPlayer 类结合，实现视频输出功能。Android 还提供了一个 SurfaceView 的子类 VideoView，通过它，可以直接在界面中建立一个用于播放视频的控件，与使用 SurfaceView 控件进行视频开发相比，VideoView 的编程更加方便，但是功能扩展性和编程自由度低于前者。

6.5.1 使用 MediaPlayer 和 SurfaceView 播放视频

SurfaceView 继承自 View，其中内嵌了一个专门用于绘制的 Surface。开发者可以控制这个 Surface 的格式和尺寸，并让 SurfaceView 控制这个 Surface 的绘制位置。Surface 对应一块屏幕缓冲区，每个 Window 对应一个 Surface，任何 View 都是画在 Surface 上的，包括之前讲过的控件和布局控件。一般的 View 都共享一块屏幕缓冲区，所有的绘制必须在 UI 线程中进行。

Surface 是纵深排序（Z-ordered）的，这表明它总在自己所在窗口（Window）的后面。SurfaceView 提供了一个可见区域，只有在这个可见区域内的 Surface 部分内容才可见，可见区域外的部分不可见。如果 Surface 上面有透明控件，那么它的每次变化都会引起对它和顶层控件之间透明关系的重新运算，而这会影响性能。

可以通过 SurfaceHolder 接口访问这个 Surface，通过 getHolder()方法可以得到这个接口。SurfaceHolder.Callback 用来实现接口，以接收 surface 变化的消息，为 Surface 设置 Callback 的方法是 SurfaceHolder.addCallback。

SurfaceHolder.Callback 的几个重要方法，说明如下。

（1）surfaceCreated(SurfaceHolder holder)：当 Surface 第一次创建后会立即调用该函数。程序可以在该函数中做些和绘制界面相关的初始化工作，一般情况下都是在另外的线程来绘制界面,所以不要在这个函数中绘制 Surface。

(2) surfaceChanged(SurfaceHolder holder, int format, int width, int height):当 Surface 的状态(大小和格式)发生变化的时候会调用该函数,在 surfaceCreated 调用后该函数至少会被调用一次。

(3) surfaceDestroyed(SurfaceHolder holder):当 Surface 被摧毁前会调用该函数,该函数被调用后就不能继续使用 Surface 了,一般在该函数中清理使用的资源。

SurfaceView 变得可见时 Surface 被创建,SurfaceView 隐藏前 Surface 被销毁,这样能节省资源。如果要监视 Surface 被创建和销毁的过程,可以实现 surfaceCreated (SurfaceHolder)和 surfaceDestroyed(SurfaceHolder)方法。

SurfaceView 的核心在于提供了两个线程:UI 线程和绘图线程(Drawing Thread)。这里应注意如下两点:

(1) 所有 SurfaceView 和 SurfaceHolder.Callback 的方法都应该在 UI 线程里调用,绘图线程所要访问的各种变量应该作同步处理。

(2) 必须确保绘图线程只能在它本身有效的情况下接触其下的 Surface,而绘图线程在 SurfaceHolder.Callback.surfaceCreated()和 SurfaceHolder.Callback.surfaceDestroyed()的调用之间是有效的。

通过 SurfaceView 与 MediaPlayer 结合的方式实现视频播放功能的步骤如下:

(1) 创建 MediaPlayer 对象,并通过 setDataSource()设置加载的视频文件;

(2) 在界面布局文件中定义 SurfaceView 控件;

(3) 通过 MediaPlayer.setDisplay(SurfaceHolder sh)来指定视频画面输出到 SurfaceView 之上;

(4) 通过 MediaPlayer 的其他一些方法播放视频。

6.5.2 使用 MediaPlayer 和 SurfaceView 播放视频案例

1. 案例功能描述

案例将提供一个简单的视频播放器,实现视频的播放、暂停和停止三个功能。

2. 案例程序结构

案例中包括一个布局文件(activity_main.xml);一个 Activity 组件类(MainActivity 类),用于实现用户界面交互功能。

3. 案例的实现步骤和思路

(1) 创建 Android 项目。

(2) 在 res 目录下的 layout 子目录中创建新的布局文件 activity_main.xml,最外层容器为 RelativeLayout 布局,宽和高都是充满父容器,内部有三个按钮控件和一个 surfaceView。

(3) 在 src 目录下创建服务类 MediaPlayoerService,继承 Service。

(4) 编写 MainActivity 文件,复写生命周期方法 onCreate。
① 在 onCreate 方法中,先调用父类的 onCreate 方法;
② 使用 setContentView 方法加载布局文件 activity_main.xml;
③ 获取界面的三个按钮控件,给每个按钮设置监听器;
④ 创建连接对象获取 SurfaceView 对象;
⑤ 使用 SurfaceView 对象展示媒体播放。

4. 案例参考代码

(1) 布局文件 activity_main.xml 代码如下。

```xml
<?xml version="1.0" encoding="utf-8"?>
<!--主界面布局,最外面是相对布局 -->
<RelativeLayout xmlns:android="http://schemas.android.com/apk/res/android"
    xmlns:tools="http://schemas.android.com/tools"
    android:layout_width="match_parent"
    android:layout_height="match_parent"
    tools:context=".VideoPlayerActivity">
    <!--布局中第一个控件 SurfaceView,用于显示视频图像 -->
    <SurfaceView
        android:id="@+id/surfaceView1"
        android:layout_width="match_parent"
        android:layout_height="150dp" />
    <!--布局中第二个控件 Button -->
    <Button
        android:id="@+id/btnPlay"
        android:layout_width="wrap_content"
        android:layout_height="wrap_content"
        android:layout_alignParentLeft="true"
        android:layout_centerVertical="true"
        android:layout_marginLeft="15dp"
        android:text="播放" />
    <!--布局中第三个控件 Button -->
    <Button
        android:id="@+id/btnPause"
        android:layout_width="wrap_content"
        android:layout_height="wrap_content"
        android:layout_alignBaseline="@+id/btnPlay"
        android:layout_alignBottom="@+id/btnPlay"
        android:layout_centerHorizontal="true"
        android:text="暂停" />
    <!--布局中第四个控件 Button -->
    <Button
        android:id="@+id/btnStop"
```

```
            android:layout_width="wrap_content"
            android:layout_height="wrap_content"
            android:layout_alignBaseline="@+id/btnPause"
            android:layout_alignBottom="@+id/btnPause"
            android:layout_alignParentRight="true"
            android:layout_marginRight="18dp"
            android:text="停止" />
</RelativeLayout>
```

(2) Activity 组件 MainActivity 类代码如下。

```
package com.example.android_demo6_4;

import android.app.Activity;
import android.media.AudioManager;
import android.media.MediaPlayer;
import android.os.Bundle;
import android.view.SurfaceHolder;
import android.view.SurfaceHolder.Callback;
import android.view.SurfaceView;
import android.view.View;
import android.view.View.OnClickListener;
import android.widget.Button;

/**
 * 主Activity为控制类,用于控制界面的显示
 *
 */
public class MainActivity extends Activity implements OnClickListener {

    private Button btnPlay, btnPause, btnStop;
    private SurfaceView surfaceView;
    private MediaPlayer player;
    private int position;

    /**
     * Activity组件的生命周期方法,在组件创建时调用,一般用于初始化信息
     */
    @SuppressWarnings("deprecation")
    @Override
    public void onCreate(Bundle savedInstanceState) {
        //调用父类方法,完成系统工作
        super.onCreate(savedInstanceState);
        //加载界面布局文件
        setContentView(R.layout.activity_main);
```

```java
//获取界面控件
btnPlay=(Button) this.findViewById(R.id.btnPlay);
btnPause=(Button) this.findViewById(R.id.btnPause);
btnStop=(Button) this.findViewById(R.id.btnStop);
btnPlay.setOnClickListener(this);
btnPause.setOnClickListener(this);
btnStop.setOnClickListener(this);
//初始化 MediaPlayer
player=new MediaPlayer();
//得到 SurfaceView 对象,它指向界面中的一块 SurfaceView 区域,用来显示图像
surfaceView=(SurfaceView) this.findViewById(R.id.surfaceView1);
//设置 SurfaceView 自己不管理的缓冲区
surfaceView.getHolder()
        .setType(SurfaceHolder.SURFACE_TYPE_PUSH_BUFFERS);
surfaceView.getHolder().addCallback(new Callback() {
    public void surfaceDestroyed(SurfaceHolder holder) {
    }
    public void surfaceCreated(SurfaceHolder holder) {
        if (position>0) {
            try {
                //开始播放
                play();
                //直接从指定位置开始播放
                player.seekTo(position);
                position=0;
            } catch (Exception e) {
            }
        }
    }
    public void surfaceChanged(SurfaceHolder holder, int format,
            int width, int height) {
    }
});
}

/**
 * 按钮单击事件的处理方法
 */
@Override
public void onClick(View v) {
    switch (v.getId()) {
    case R.id.btnPlay:
        play();
        break;
```

```java
            case R.id.btnPause:
                if (player.isPlaying()) {
                    player.pause();
                } else {
                    player.start();
                }
                break;
            case R.id.btnStop:
                if (player.isPlaying()) {
                    player.stop();
                }
                break;
            default:
                break;
        }
    }

    /**
     * 暂停方法
     */
    protected void onPause() {
        //先判断是否正在播放
        if (player.isPlaying()) {
            //如果正在播放就先保存这个播放位置
            position=player.getCurrentPosition();
            player.stop();
        }
        super.onPause();
    }

    /**
     * 播放方法
     */
    private void play() {
        try {
            player.reset();
            player.setAudioStreamType(AudioManager.STREAM_MUSIC);
            //设置需要播放的视频,事先将其复制到设备的 SD 卡中,这里直接放在 SD 卡的根
            //目录下
            player.setDataSource("/mnt/sdcard/video.avi");
            //把视频画面输出到 SurfaceView
            player.setDisplay(surfaceView.getHolder());
            player.prepare();
            //播放
```

```
            player.start();
        } catch (Exception e) {
        }
    }
}
```

5. 案例运行效果

使用 MediaPlayer 播放视频的运行效果如图 6-5 所示。

图 6-5 MediaPlayer 播放视频运行效果

6.5.3 使用 VideoView 播放视频

　　VideoView 是 Android 提供的一个媒体播放显示和控制的控件,它继承自 SurfaceView。在使用 VideoView 时,不需要管理 MediaPalyer 的各种状态,因为这些状态都已经被 VideoView 封装好了。当 VideoView 被创建的时候,MediaPalyer 对象将会被创建;当 VideoView 被销毁的时候,MediaPlayer 对象也随之被释放。

　　一个使用 VideoView 播放视频文件的代码片段如下所示。

```
protected void onCreate(Bundle savedInstanceState) {
    super.onCreate(savedInstanceState);
    setContentView(R.layout.activity_main);
    video1=(VideoView)findViewById(R.id.video1);
    mediaco=new MediaController(this);
    File file=new File("/mnt/ext_sdcard/DCIM/Camera/ok1.mp4");
    if(file.exists()){
        video1.setVideoPath(file.getAbsolutePath());
        //VideoView 与 MediaController 进行关联
        video1.setMediaController(mediaco);
```

```
        mediaco.setMediaPlayer(video1);
        //让 VideiView 获取焦点
        video1.requestFocus();
    }
}
```

在上面的代码中,MediaController 类主要实现对 VideoView 的播放控制,它是一个包含媒体播放器(MediaPlayer)的媒体控制条。MediaController 与 VideoView 配合使用,能实现播放界面的主要功能。上面代码的运行效果如图 6-6 所示。

图 6-6 VideoView 播放视频运行效果

6.6 实现拍照功能

Android 提供了两种方法来实现拍照功能,一是借助 Intent 和 MediaStroe 调用系统自带的拍照应用程序来实现拍照功能;二是使用 Camera 类中的 API 自行编写拍照程序。

6.6.1 使用系统自带的拍照应用程序

由于 Android 系统自带了拍照应用,且能被其他应用调用,所以可以直接调用系统自带的拍照应用,来实现拍照功能。如下代码片段实现了调用 Android 系统自带的拍照应用进行拍照,拍照完成后会将照片保存在目录/sdcard/pics 中,文件名为 pic01.jpg。

```
//保存所拍照片的文件路径,必须确保文件夹路径存在,否则拍照后无法完成回调
String imgPath="/sdcard/pics/pic01.jpg";
File file=new File(imgPath);
if (!file.exists()) {
    File path=file.getParentFile();
    path.mkdirs();
}
Uri uri=Uri.fromFile(file);
Intent intent=new Intent(MediaStore.ACTION_IMAGE_CAPTURE);
```

```
intent.putExtra(MediaStore.EXTRA_OUTPUT, uri);
startActivity(intent);
```

6.6.2 自行开发拍照功能

由于系统自带的拍照应用功能固定且界面单一,所以如果想做出功能丰富且界面绚丽的拍照应用,就必须使用 Camera 类中的 API 来自行开发。

调用系统自带的拍照应用实现拍照功能,不需要加入任何的权限,但是对于调用 Camera 类的 API 自行开发拍照应用的方式,则需要引入下面的几个权限:

```
<uses-permission android:name="android.permission.CAMERA" />
<uses-feature android:name="android.hardware.camera" android:required="true"/>
<uses-feature android:name="android.hardware.camera.autofocus"
android:required="true"/>
```

一般都把拍照后得到的图片存放在 SD Card 中,所以通常还需要声明对外部存储的写权限:

```
<uses-permission android:name="android.permission.WRITE_EXTERNAL_STORAGE"/>
```

由于与图像有关,所以拍照功能中也用到了 SurfaceView 控件,使用 Cemera 类的 API 开发拍照应用的大致步骤如下。

(1) 在 Activity 的 onCreate()方法中设置 SurfaceView 对象。

(2) 在 SurfaceHolder.Callback 的 surfaceCreated()中,调用 Camera 的 open()方法;之后调用 getParameters()方法得到已打开的摄像头的配置参数 Parameters 对象;通过 Parameters 对象给 Camera 设置一系列参数;调用 Camera 对象的 startPreview()方法,开始拍照预览。

(3) 调用 takePicture(Camera.ShutterCallback,Camera.PictureCallback,Camera.PictureCallback,Camera.PictureCallback)方法完成拍照。

上面的步骤只是说明了实现拍照功能的大致流程,下面将通过一个案例来展示如何一步步地实现拍照功能。

6.6.3 Camera 类使用案例

案例使用 Camera 类和 SurfaceView 控件实现了简单的拍照功能。案例程序包括一个 Activity 类(MainActivity.java)及其界面布局文件(activity_main.xml),一个拍照预览类(Preview.java)。

1. 入口 Activity 代码 CameraActivity.java

```
package com.example.android_demo6_6;

import android.app.Activity;
import android.os.Bundle;
```

```java
import android.util.Log;
import android.view.Menu;
import android.view.MenuItem;
import android.view.View;
import android.widget.FrameLayout;
import android.widget.ImageButton;
public class MainActivity extends Activity {
    FrameLayout fl;
    ImageButton btn;
    Preview preview;

    @Override
    protected void onCreate(Bundle savedInstanceState) {
        super.onCreate(savedInstanceState);
        setContentView(R.layout.activity_main);
        fl= (FrameLayout) findViewById(R.id.wPreview);
        preview=new Preview(this);
        fl.addView(preview);
        btn= (ImageButton) findViewById(R.id.imageButton1);
        btn.setOnClickListener(new View.OnClickListener() {
            @Override
            public void onClick(View v) {
                //拍照
                preview.tackPicture();
            }
        });
    }
}
```

2. 拍照预览类 Preview.java

```java
package com.example.android_demo6_6;

import java.io.FileOutputStream;
import java.io.IOException;
import java.io.OutputStream;
import java.util.List;
import android.content.Context;
import android.graphics.Bitmap;
import android.graphics.BitmapFactory;
import android.graphics.PixelFormat;
import android.hardware.Camera;
import android.hardware.Camera.Size;
```

```java
import android.os.Environment;
import android.util.Log;
import android.view.SurfaceHolder;
import android.view.SurfaceView;

@SuppressWarnings("deprecation")
public class Preview extends SurfaceView implements SurfaceHolder.Callback,
        Camera.PictureCallback {
    SurfaceHolder surfaceHolder;
    Camera camera;
    Context context;
    Camera.Parameters parameters;
    int maxZoom, currentZoom;

    @SuppressWarnings("deprecation")
    Preview(Context context) {
        super(context);
        this.context=context;
        /*取得 holder*/
        surfaceHolder=getHolder();
        surfaceHolder.addCallback(this);
        /*设定预览 Buffer Type*/
        surfaceHolder.setType(SurfaceHolder.SURFACE_TYPE_PUSH_BUFFERS);
    }

    @SuppressWarnings("deprecation")
    @Override
    public void surfaceCreated(SurfaceHolder holder) {

        /*若相机在非预览模式,则开启相机*/
        camera=Camera.open();
        try {
            camera.setPreviewDisplay(holder);
        } catch (IOException e) {
            //TODO Auto-generated catch block
            e.printStackTrace();
        }
        camera.setPreviewCallback(new Camera.PreviewCallback() {
            public void onPreviewFrame(byte[] data, Camera arg1) {
                /*在此可针对预览图像做一些优化*/
            }
        });
```

```java
        /*建立 Camera.Parameters*/
        parameters=camera.getParameters();
        List<Size>sizes=parameters.getSupportedPreviewSizes();
        for (Size size : sizes) {
            Log.e("SupportedPreviewSizes:", size.width+" * "+size.height);
        }
        /*设定预览画面大小*/
        parameters.setPreviewSize(640, 480);
        List<String>focusModes=parameters.getSupportedFocusModes();
        for (String focusMode : focusModes) {
            Log.e("SupportedFocusModes:", focusMode);
        }
        /*设定对焦模式*/
        parameters.setFocusMode (Camera.Parameters.FOCUS_MODE_CONTINUOUS_PICTURE);
        /*设定图像格式*/
        parameters.setPictureFormat(PixelFormat.JPEG);
        /*将上述设定参数给 Camera*/
        camera.setParameters(parameters);
        camera.setDisplayOrientation(90);
        /*立即执行 Preview*/
        camera.startPreview();
    }

    @Override
    public void surfaceDestroyed(SurfaceHolder holder) {
        /*停止 Preview*/
        camera.stopPreview();
        camera.release();
        camera=null;
    }

    @Override
    public void surfaceChanged(SurfaceHolder holder,int format,int w,int h) {
        System.out.println("surfaceChanged call.");
    }

    public void tackPicture() {
        camera.takePicture(null, null, null, this);
    }

    @Override
    public void onPictureTaken(byte[] data, Camera camera) {
```

```
            try {
                OutputStream os=new FileOutputStream(Environment
                    .getExternalStorageDirectory().getAbsolutePath()+"/"
                    +System.currentTimeMillis()+".jpg");
                Bitmap bmp=BitmapFactory.decodeByteArray(data, 0, data.length);
                bmp.compress(Bitmap.CompressFormat.JPEG, 100, os);
                os.close();
            } catch (Exception e) {
                e.printStackTrace();
            }
            camera.startPreview();
        }
}
```

3. 入口 Activity 的界面布局文件 camera.xml

```
<?xml version="1.0" encoding="utf-8"?>
<RelativeLayout xmlns:android="http://schemas.android.com/apk/res/android"
    android:id="@+id/layout"
    android:layout_width="fill_parent"
    android:layout_height="fill_parent">

    <FrameLayout
        android:id="@+id/wPreview"
        android:layout_width="match_parent"
        android:layout_height="match_parent"
        android:layout_margin="10dp">

    </FrameLayout>

    <ImageButton
        android:id="@+id/imageButton1"
        android:layout_width="50dp"
        android:layout_height="50dp"
        android:layout_alignParentBottom="true"
        android:layout_centerHorizontal="true"
        android:layout_margin="20dp"
        android:background="@drawable/but_p1" />

</RelativeLayout>
```

案例的运行效果如图 6-7 所示。

图 6-7 使用 Camera 类实现的拍照功能

6.7 Android 动画设计

Android 中的动画可以分为两种,一种是逐帧(Frame)动画,另一种是补间(Tween)动画。逐帧动画很容易理解,要求开发者把动画过程的每一张图片都收集起来,然后由 Android 来控制和显示这些静态图片,利用人眼"视觉暂留"的原理,给用户创造动画的感觉。逐帧动画像放电影一样,每秒内放多少张胶片,让人感觉图像一直在动。补间动画只需要定义动画开始、动画结束的"关键帧",动画变化的"中间帧"由 Android 系统计算、补齐。

6.7.1 Android 中的逐帧动画

逐帧动画是以很短的间隔连续显示一系列图像的简单过程,所以最终效果是一个移动或者变化的对象。

在开发逐帧动画之前,首先需要使用一系列图像来计划动画顺序。图 6-8 展示了一组大小相同的小木人,每一个小木人的大小一致,但是有一些动作变化。像这样的图片保存十几张之后,就可以使用动画来展示小木人的移动了。

下面将通过一个案例说明 Android 中的逐帧动画是如何实现的。

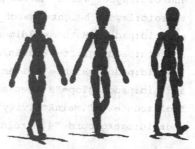

图 6-8 展示小木人移动的一组图

6.7.2 逐帧动画演示案例

案例将实现一个小人的移动动画,通过单击界面上的按钮,可以让动画开始或停止。案例程序包含一个逐帧动画文件(frame_animation.xml),一个 Activity 类(MainActivity.java)及其界面布局文件(activity_main.xml)。

1. drawable 文件夹中的逐帧动画文件 frame_animation.xml

```
<?xml version="1.0"encoding="utf-8"?>
<animation-list
xmlns:android="http://schemas.android.com/apk/res/android"
android:oneshot="false"><item
android:drawable="@drawable/p1" android:duration="60"/>
<item
android:drawable="@drawable/p2" android:duration="60"/>

<!-中间省略图像 p3 到 p29 的代码 -->

<item
```

```
android:drawable="@drawable/p30" android:duration="60"/>
</animation-list>
```

上述代码中通过<animation-list>标签来表示一个动画集合,在集合中可以通过 android:oneshot="[true|false]"为动画设置是否循环播放;每一个<item>标签代表一帧,其中利用 android:drawable 属性为这一帧设置图片;android:duration 属性表示这一帧所持续的时间,单位为 ms。

当设置好动画的 XML 文件之后,在界面的布局中可以为视图(ImageView)组件设置动画资源,可以通过 android:src 属性指定。

2. 入口 Activity 布局文件 activity_main.xml

```xml
<RelativeLayout
xmlns:android="http://schemas.android.com/apk/res/android"
xmlns:tools="http://schemas.android.com/tools"
android:layout_width="match_parent"
android:layout_height="match_parent"
android:paddingBottom="@dimen/activity_vertical_margin"
android:paddingLeft="@dimen/activity_horizontal_margin"
android:paddingRight="@dimen/activity_horizontal_margin"
android:paddingTop="@dimen/activity_vertical_margin"
tools:context=".MainActivity"
android:background="@android:color/black">

    <ImageView
        android:id="@+id/imageView1"
        android:layout_width="wrap_content"
        android:layout_height="wrap_content"
        android:layout_alignParentTop="true"
        android:layout_centerHorizontal="true"
        android:layout_marginTop="45dp"
        android:src="@drawable/frame_animation"/>

    <Button
        android:id="@+id/button1"
        android:layout_width="wrap_content"
        android:layout_height="wrap_content"
        android:layout_below="@+id/imageView1"
        android:layout_centerHorizontal="true"
        android:layout_marginTop="30dp"
        android:text="StartAnimation"/>

    <Button
        android:id="@+id/button2"
```

```
                android:layout_width="wrap_content"
                android:layout_height="wrap_content"
                android:layout_below="@+id/button1"
                android:layout_centerHorizontal="true"
                android:layout_marginTop="20dp"
                android:text="StopAnimation"/>

</RelativeLayout>
```

3. 入口 Activity 类 MainActivity.java

```java
package com.example.android_demo6_7;

import android.app.Activity;
import android.graphics.drawable.AnimationDrawable;
import android.os.Bundle;
import android.view.Menu;
import android.view.View;
import android.view.View.OnClickListener;
import android.widget.Button;
import android.widget.ImageView;

public class MainActivity extends Activity {

    private ImageView iv;
    private Button startBtn, stopBtn;
    //声明动画
    private AnimationDrawable anim;

    private void setupView(){
        iv=(ImageView) this.findViewById(R.id.imageView1);
        startBtn=(Button) this.findViewById(R.id.button1);
        stopBtn=(Button) this.findViewById(R.id.button2);
        anim=(AnimationDrawable)iv.getDrawable();

    }

    private void addListener(){
        startBtn.setOnClickListener(new OnClickListener() {

            @Override
            public void onClick(View v) {
                //TODO Auto-generated method stub
```

```
            anim.start();
        }
    });
    stopBtn.setOnClickListener(new OnClickListener() {
        
        @Override
        public void onClick(View v) {
            //TODO Auto-generated method stub
            anim.stop();
        }
    });
}

@Override
protected void onCreate(Bundle savedInstanceState) {
    super.onCreate(savedInstanceState);
    setContentView(R.layout.activity_main);
    //引用初始化方法
    setupView();
    //引用添加监听器方法
    addListener();
}

@Override
public boolean onCreateOptionsMenu(Menu menu) {
    //Inflate the menu; this adds items to the action bar if it is present.
    getMenuInflater().inflate(R.menu.main, menu);
    returntrue;
}

}
```

在上面的代码中,通过 findViewById()方法找到对应的 ImageView 组件,并通过 ImageView 组件中的 getDrawable()方法获得 AnimationDrawable 对象,然后使用该对象中的 start()方法开启动画,如果需要结束动画可以调用 stop()方法。案例运行效果如图 6-9 所示。

6.7.3 Android 中的补间动画

与逐帧动画不同,补间动画不是通过重复帧来实现的,而是通过不断地改变视图的属性来实现的。

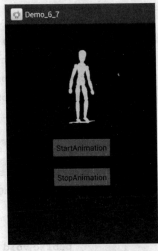

图 6-9 逐帧动画运行效果

1. Alpha(透明度)动画

创建透明度动画时要指定动画开始时透明度、结束时透明度及动画的持续时间。透明度的取值是 0.0~1.0 之间。透明度为 1.0 代表完全不透明，0.0 代表完全透明。参考代码片段如下。

```
<?xml version="1.0"encoding="utf-8"?>
<alpha
xmlns:android="http://schemas.android.com/apk/res/android"
android:fromAlpha="1"
android:toAlpha="0.5"
android:duration="5000"
android:fillAfter="true"
/>
```

上述代码中，android:fromAlpha 属性为起始透明度，android:toAlpha 属性为结束透明度，android:duration 属性为动画的持续时间(单位为 ms)，android:fillAfter 属性代表动画是否停留在动画结束的位置上。

2. Scale(缩放)动画

创建缩放动画效果时需要指定动画开始时的缩放比例、结束时的缩放比例，并指定动画持续的时间。由于缩放动画以不同基准点缩放的效果不同，因此还需要指定缩放动画的缩放中心点。参考代码片段如下。

```
<?xml version="1.0"encoding="utf-8"?>
<scale xmlns:android="http://schemas.android.com/apk/res/android"
android:fromXScale="1"
android:toXScale="0.5"
android:fromYScale="1"
android:toYScale="0.5"
android:pivotX="50%"
android:pivotY="50%"
android:duration="5000"
android:interpolator="@android:anim/accelerate_decelerate_interpolator"
/>
```

上述代码中 android:fromXScale 属性和 android:fromYScale 属性用来设置开始动画的 X 轴和 Y 轴的缩放比例，android:toXScale 属性和 android:toYScale 属性用来设置结束动画时 X 轴和 Y 轴的缩放比例，android:pivotX 属性和 android:pivotY 属性用来设置缩放中心点，以百分比来表示。而 android:interpolator 属性用来指定一个动画的插入器。它的常用参数如下。

(1) accelerate_decelerate_interpolator：先加速后减速动画；

(2) accelerate_interpolator：减速动画；

(3) decelerate_interpolator：加速动画。

3. Translate（位）移动画

创建位移动画时需要指定动画开始时的位置、结束时的位置以及动画持续的时间。参考代码片段如下。

```xml
<?xml version="1.0"encoding="utf-8"?>
<translate
xmlns:android="http://schemas.android.com/apk/res/android"
android:fromXDelta="10"
android:toXDelta="110"
android:fromYDelta="10"
android:toYDelta="110"
android:duration="5000"/>
```

上述代码中 android:fromXDelta 属性和 android:fromYDelta 属性为动画起始时的 X 轴和 Y 轴的位置坐标，而 android:toXDelta 属性和 android:toYDelta 属性为动画结束时的 X 轴和 Y 轴的位置坐标。

4. Rotate（旋转）动画

创建旋转动画时需要指定动画开始时的旋转角度、结束时的旋转角度、动画持续的时间。由于旋转动画以不同的点为中心旋转的效果不同，因此还需要知道指定"旋转轴心"的坐标。参考代码片段如下。

```xml
<?xml version="1.0"encoding="utf-8"?>
<rotate
xmlns:android="http://schemas.android.com/apk/res/android"
android:fromDegrees="0"
android:toDegrees="-360"
android:pivotX="50%"
android:pivotY="50%"
android:duration="5000"/>
```

上述代码中 android:fromDegrees 属性为起始角度，android:toDegrees 属性为结束角度。当角度为正数时动画顺时针旋转，角度为负数时动画逆时针旋转。

6.7.4 补间动画演示案例

案例将实现 6.7.3 节介绍的几种动画效果。案例程序包括补间动画文件 tween_animation.xml，一个 Activity 类（MainActivity.java）及其界面布局文件（activity_main.xml）。在 Activity 类中可以通过 AnimationUtils 中的 loadAnimation() 方法加载 XML 动画资源，获得 Animation 动画对象，然后调用 startAnimation() 方法来开启动画。

1. anim 文件夹下的补间动画文件 tween_animation.xml

```xml
<?xml version="1.0"encoding="utf-8"?>
<set xmlns:android="http://schemas.android.com/apk/res/android">
<alpha
android:fromAlpha="0.5"
android:toAlpha="1"
android:duration="5000"/>
<rotate
android:fromDegrees="0"
android:toDegrees="360"
android:pivotX="50%"
android:pivotY="50%"
android:duration="5000"/>
<scale
android:fromXScale="0.5"
android:toXScale="1"
android:fromYScale="0.5"
android:toYScale="1"
android:pivotX="50%"
android:pivotY="50%"
android:duration="5000"/>
</set>
```

2. 入口 Activity 类 MainActivity.java

```java
package com.example.android_demo6_4;

import android.app.Activity;
import android.os.Bundle;
import android.view.animation.Animation;
import android.view.animation.AnimationUtils;
import android.widget.ImageView;
import android.widget.RadioGroup;
import android.widget.RadioGroup.OnCheckedChangeListener;

public class MainActivity extends Activity {

    //声明控件
    private RadioGroup rg;
    private ImageView iv;

    @Override
    protected void onCreate(Bundle savedInstanceState) {
```

```java
        super.onCreate(savedInstanceState);
        setContentView(R.layout.activity_main);
        //初始化方法
        setupView();
        //引用添加监听器方法
        addListener();
    }

    /**
     * 初始化控件方法
     */
    private void setupView() {
        rg=(RadioGroup) this.findViewById(R.id.rg);
        iv=(ImageView) this.findViewById(R.id.imageView1);
    }

    /**
     * 添加监听器方法
     */
    private void addListener() {
        //RadioGroup 添加监听器
        rg.setOnCheckedChangeListener(new OnCheckedChangeListener() {

            @Override
            public void onCheckedChanged(RadioGroup group, int checkedId) {
                //RadioButton 判断单击
                switch (checkedId) {
                case R.id.rb1:
                    //获得动画对象
                    Animation anim1=AnimationUtils.loadAnimation(
                            MainActivity.this, R.anim.alpha_animation);
                    //启动动画
                    iv.startAnimation(anim1);
                    break;
                case R.id.rb2:
                    Animation anim2=AnimationUtils.loadAnimation(
                            MainActivity.this, R.anim.rotate_animation);
                    iv.startAnimation(anim2);
                    break;
                case R.id.rb3:
                    Animation anim3=AnimationUtils.loadAnimation(
                            MainActivity.this, R.anim.scale_animation);
                    iv.startAnimation(anim3);
                    break;
```

```java
                case R.id.rb4:
                    Animation anim4=AnimationUtils.loadAnimation(
                            MainActivity.this, R.anim.translate_animation);
                    iv.startAnimation(anim4);
                    break;
                case R.id.rb5:
                    Animation anim5=AnimationUtils.loadAnimation(
                            MainActivity.this, R.anim.tween_animation);
                    iv.startAnimation(anim5);
                    break;
                default:
                    break;
                }
            }
        });
    }
}
```

3. 入口 Activity 布局文件 activity_main.xml

```xml
<RelativeLayout
xmlns:android="http://schemas.android.com/apk/res/android"
xmlns:tools="http://schemas.android.com/tools"
android:layout_width="match_parent"
android:layout_height="match_parent"
android:paddingBottom="@dimen/activity_vertical_margin"
android:paddingLeft="@dimen/activity_horizontal_margin"
android:paddingRight="@dimen/activity_horizontal_margin"
android:paddingTop="@dimen/activity_vertical_margin"
tools:context=".MainActivity">
<RadioGroup
android:id="@+id/rg"
android:layout_width="fill_parent"
android:layout_height="wrap_content"
android:orientation="vertical">
<RadioButton
android:id="@+id/rb1"
android:layout_width="wrap_content"
android:layout_height="wrap_content"
android:text="透明度"/>
<RadioButton
android:id="@+id/rb2"
android:layout_width="wrap_content"
```

```
        android:layout_height="wrap_content"
        android:text="旋转"/>
    <RadioButton
        android:id="@+id/rb3"
        android:layout_width="wrap_content"
        android:layout_height="wrap_content"
        android:text="缩放"/>
    <RadioButton
        android:id="@+id/rb4"
        android:layout_width="wrap_content"
        android:layout_height="wrap_content"
        android:text="位移"/>
    <RadioButton
        android:id="@+id/rb5"
        android:layout_width="wrap_content"
        android:layout_height="wrap_content"
        android:text="综合"/>
    </RadioGroup>
    <ImageView
        android:id="@+id/imageView1"
        android:layout_width="wrap_content"
        android:layout_height="wrap_content"
        android:layout_centerInParent="true"
        android:background="@drawable/ben_window_0001" />
</RelativeLayout>
```

案例运行效果，如图 6-10 所示。

6.7.5 动画监听事件

动画在执行过程中，如果需要监听动画状态，可以使用 Android 提供的接口 AnimationListener 实现，该接口中需要实现其内部的如下三个方法。

（1）onAnimationStart（Animation animation）：动画开始时回调；

（2）onAnimationRepeat（Animation animation）：动画进行时回调；

（3）onAnimationStart（Animation animation）：动画结束时回调。

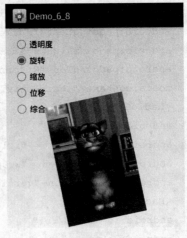

图 6-10 补间动画运行结果

下面的代码片段实现了补间动画的监听功能，即在 Activity 类中实现 AnimationListener 接口，通过 setAnimationListener（）为补间动画设置了监听器，当动画结束时，会出现提示"动画结束"。

```java
public class MainActivity extends Activity implements AnimationListener{
    private ImageView iv;
    private Animation anim;
    @Override
    protected void onCreate(Bundle savedInstanceState) {
        super.onCreate(savedInstanceState);
        setContentView(R.layout.activity_main);
        iv= (ImageView) this.findViewById(R.id.iv);
        anim=new AlphaAnimation(0.5f, 1.0f);
        anim.setDuration(5000);
        //开启动画
        iv.startAnimation(anim);
        //为补间动画设置监听器
        anim.setAnimationListener(this);
    }
    @Override
    public void onAnimationStart(Animation animation) {
        //TODO Auto-generated method stub
    }
    @Override
    public void onAnimationEnd(Animation animation) {
        Toast.makeText(this, "动画结束", Toast.LENGTH_LONG).show();
    }
    @Override
    public void onAnimationRepeat(Animation animation) {
        //TODO Auto-generated method stub
    }
}
```

习 题 6

1. Android 中如何动态绘制图形？
2. 简述 MediaPlayer 的生命周期。
3. 逐帧(Frame)动画与补间(Tween)动画有什么区别，一般该如何实现？

第 7 章

Android 传感器

在 Android 中,传感器是一个集成在设备中的硬件,可以从物理环境中获得数据提供给应用程序,应用程序需要传感器数据传递给用户,完成一定的操作或者显示环境数据,例如控制游戏、显示周围的温度等。传感器设备是只读的,需要设置监听器来接受传感器数据。在 Android 传感器框架内可以访问多种类型的传感器。这些传感器有的是基于硬件的,有的是基于软件的。基于硬件的传感器是内置到手机或平板设备的物理组件,它们通过直接测量获得特定的环境数据,如加速度、地磁场的强度或角度变化。基于软件的传感器没有具体的物理设备,而是模仿了硬件传感器。基于软件的传感器是从一个或更多的基于硬件的传感器获得数据,有时我们称其为虚拟传感器或合成传感器。例如,线性加速度计和重力传感器是基于软件的传感器。

7.1 传感器的分类

根据传感器获取数据的种类,Android 平台提供了三类传感器类型,但是并不是每部手机都集成了所有的传感器。

7.1.1 移动传感器

移动传感器是度量设备在三个轴上的加速度和旋转角度,其中包括加速度(TYPE_ACCELEROMETER)、重力(TYPE_GRAVITY)、陀螺仪(TYPE_GYROSCOPE)、线性加速(TYPE_LINEAR_ACCELERATION)和旋转矢量(TYPE_ROTATION_VECTOR)传感器。

Android 平台支持很多监测设备运动的传感器。其中有两个传感器一定是基于硬件的,例如加速度计和陀螺仪,还有三个可能基于硬件或者软件,例如重力、线性加速计和旋转矢量传感器。例如,某些设备的软传感器利用加速度计和磁力计来获得它们的数据,而其他一些设备可能用陀螺仪来获得它们的数据。大部分 Android 平台的设备都带有加速度计,有很多现在还带有陀螺仪。软传感器的可用性变数更大一些,因为它们常常依靠一个以上硬件传感器来报送数据。

移动传感器对于监测设备的运动非常有用,例如倾斜、震动、旋转、摇摆等。这些动作通常是直观反映了用户的输入(例如用户在游戏中操纵汽车或者运球),但也可能反映了设备所处的物理环境变化(例如在开车时,设备也随着移动)。

7.1.2 位置传感器

Android 平台提供了两种传感器来检测设备的方位:地磁场传感器和方向传感器。Android 平台还提供了一种传感器,用于检测屏幕表面与其他物体的邻近程度,即被称为距离传感器。地磁场传感器和距离传感器是基于硬件的。

大部分手持和桌面设备都内置了地磁传感器。手持设备通常还内置了距离传感器,用于检测与人脸的靠近程度,例如识别手机与人耳的距离。而方向传感器是基于软件的,其数据来自加速度计和地磁场传感器。从 Android 2.2(API Level 8)开始,方向传感器开始被废弃了。位置传感器用于确定设备相对地球的物理方位。例如,可以用地磁传感器和加速度计来确定设备相对北极点的方位以及设备相对用户参照系的方位。位置传感器通常不会用于监测设备的移动情况,诸如震动、倾斜、冲击。地磁传感器和方向传感器在 SensorEvent 中返回以多维数组表示的传感器数据。

7.1.3 环境传感器

环境传感器是度量不同的环境参数,例如周围空气温度、压力、光线和湿度,其中包括环境温度表(TYPE_AMBIENT_TEMPERATURE)、气压力表(TYPE_PRESSURE)、测光表(TYPE_LIGHT)传感器。

Android 平台提供了 4 种用于监测环境参数的传感器,可以用这些传感器来监测 Android 设备周边环境的湿度、光照度、气压和气温。这 4 种传感器都是基于硬件的,它们都需要制造商植入设备后才能使用。目前除了制造商用于控制屏幕亮度的光线传感器以外,其他环境传感器都不一定会内置于设备中的。因此,在试图读取数据之前,验证传感器的存在性尤为重要。

由于一款 Android 设备不一定集成上面所有的传感器,许多设备只有几种,而且不同版本的 Android SDK 对传感器支持的类型也不同。那么如何知晓在 Android 设备中哪个传感器可用呢?可以通过 SensorManager(系统的传感器管理服务)获得可用的传感器对象,然后为传感器对象设置监听器来获得数据。这种方式假设用户已经安装了应用程序,但是如果没有安装怎么办呢?这时就会用到间接方式。这种方式是,在 AndroidManifest.xml 中,必须指定 Android 设备用来支持应用程序所具有的功能。例如,如果应用程序需要距离(proximity)传感器,需要在 AndroidManifest.xml 文件中添加一条内容:

```
<uses-feature android:name="android.hardware.sensor.proximity"
android:required="true" />
```

现在这个应用只能安装在具有距离传感器的 Android 设备上了。如果设置 android:required="false",这样应用程序也可以安装在没有指定传感器的设备上。

使用这种方式只能指定这个应用所需要的传感器,但是我们还是不知道这个设备上实际存在的传感器,可以使用一个简单的例子来查询设备上传感器的信息。由于会获得多个传感器,可能不在同一屏上显示,所以这个例子中使用了 ScrollView 控件。在 onCreate()方法中,可以获得 SensorManager 的引用。SensorManager 是 Android 提供的系统服务之一,所以使用了 Content 的 getSystemService()方法,如:

```
SensorManager mgr=(SensorManager) this.getSystemService(SENSOR_SERVICE);
```

然后使用 SensorManager 的 getSensorList()方法获得可用传感器的列表:

```
List<Sensor>sensors=mgr.getSensorList(Sensor.TYPE_ALL);
```

如果需要列出指定类型的传感器,可以使用其他常量代替 TYPE_ALL,例如 TYPE_GYROSCOPE、TYPE_LINEAR_ACCELERATION 或者 TYPE_GRAVITY。如果想确定设备上是否有某种传感器,可以使用 getDefaultSensor()方法。如果一种类型的传感器有多个,则必须指定一个默认的传感器;如果没有给定类型的缺省传感器,getDefaultSensor()方法的返回值为 null,例如:

```
if (mSensorManager.getDefaultSensor(Sensor.TYPE_MAGNETIC_FIELD) !=null){
    //Success! There's a magnetometer
} else {
    //Failure! No magnetometer
}
```

传感器的类型值表示了传感器的基本类型,但是对于同一基本类型的传感器来说,每个设备上的参数可能也不一样。例如光线传感器的分辨率在每个设备上是不一样的。使用<uses-feature>标签只能确定应用程序需要传感器的基本类型,如果要想确定高级的传感器参数,需要通过编写代码。其中,分辨率和最大范围就是传感器的参数;而电量以 mA 为单位,是指需要电池提供的电流,当然电量越小越好。可以使用 Sensor 类提供的公共方法获取这些参数,例如 getResolution()和 getMaximumRange()方法可以获得传感器的分辨率和最大范围,而 getPower()方法获得传感器所需要的电量。如果想根据传感器的生产厂商和版本优化我们的应用程序,还有两个方法是很有用的,分别是 getVendor()和 getVersion(),例如:

```
message.append("Vendor: "+sensor.getVendor()+"\n");
message.append("Version: "+sensor.getVersion()+"\n");
message.append("Resolution: "+sensor.getResolution()+"\n");
message.append("Max Range: "+sensor.getMaximumRange()+"\n");
message.append("Power: "+sensor.getPower()+" mA\n");
```

另一个有用的方法是 getMinDelay(),用于返回传感器采集数据的最小时间间隔(μm)。任何 getMinDelay()返回非零值的传感器都是流式传感器。流式传感器以一定的时间间隔有规律地测量数据,自 Android 2.3(API Level 9)开始引入。如果调用 getMinDelay()时返回零,这就表示该传感器不是流式传感器,只有所监测的参数发生变化时它才会报送数据。getMinDelay()方法能让我们确定传感器的最大采样频率,因此它是非常有用的。如果应用中某项功能需要很高的数据采样率或者要用到流式传感器，就可以用此方法先确认传感器是否符合要求，然后再来启用或禁用相关的功能。

目前，每种设备上集成的传感器是不同的，而且不同的 Android 版本之间也有所不同。这是因为 Android 传感器的引入经历了多个平台版本发布过程。例如，在 Android 1.5（API 等级 3）版本中，许多传感器被引入。但直到 Android 2.3（API 等级 9）这些传感器才被实现。同样，有些传感器在 Android 2.3（API 等级 9）和 Android 4.0（API 等级 14）中被引入，而有两个传感器已被弃用，取而代之的是新的、更好的传感器。表 7-1 根据 Android 平台版本总结了每种传感器的可用性。我们看到只有 4 个平台，这是因为这些平台涉及传感器的变化。被列为废弃的传感器仍然可以在随后的平台中存在，这是 Android 的向前兼容性策略。

表 7-1 Android 平台各个版本传感器的可用性

Sensor	Android 4.0 API Level 14	Android 2.3 API Level 9	Android 2.2 API Level 8	Android 1.5 API Level 3
TYPE_ACCELEROMETER	Yes	Yes	Yes	Yes
TYPE_AMBIENT_TEMPERATURE	Yes	n/a	n/a	n/a
TYPE_GRAVITY	Yes	Yes	n/a	n/a
TYPE_GYROSCOPE	Yes	Yes	n/a1	n/a1
TYPE_LIGHT	Yes	Yes	Yes	Yes
TYPE_LINEAR_ACCELERATION	Yes	Yes	n/a	n/a
TYPE_MAGNETIC_FIELD	Yes	Yes	Yes	Yes
TYPE_ORIENTATION	Yes2	Yes2	Yes2	Yes
TYPE_PRESSURE	Yes	Yes	n/a1	n/a1
TYPE_PROXIMITY	Yes	Yes	Yes	Yes
TYPE_RELATIVE_HUMIDITY	Yes	n/a	n/a	n/a
TYPE_ROTATION_VECTOR	Yes	Yes	n/a	n/a
TYPE_TEMPERATURE	Yes2	Yes	Yes	Yes

其中 n/a 1 代表的传感器类型在 Android 1.5（API Level 3）中被增加，但是直到 Android 2.3（API Level 9）中才实现；Yes 2 代表这个传感器是可用的，但是已经过时了。

7.2 获取传感器事件

当为传感器设置了监听器后，传感器就可以为应用程序提供数据；但是当传感器工作时，就会耗费电量。为保持电池寿命，需要考虑只有在需要的时候开启监听器，而在不需要时关闭监听器。怎样设置传感器的监听器呢？基本步骤如下。

（1）在 onCreate() 方法中得到 SensorManager 和 Sensor 对象，获取传感器；

(2) 在 onResume()和 onPause()回调方法中分别注册和注销传感器事件监听器；

(3) 实现 SensorEventListener 监听器接口，覆盖 onAccuracyChanged()和 onSensorChanged()两个回调方法。

假定我们通过光线传感器来测量周围的光线级别，首先需要在 onCreate()方法中得到 SensorManager 和 Sensor 对象，然后获得光线传感器，代码如下：

```java
private SensorManager mgr;
private Sensor light;
private TextView text;
private StringBuilder msg=new StringBuilder(2048);

@Override
public void onCreate(Bundle savedInstanceState) {
    super.onCreate(savedInstanceState);
    setContentView(R.layout.c08_light_sensor_layout);
    mgr= (SensorManager) this.getSystemService(SENSOR_SERVICE);
    light=mgr.getDefaultSensor(Sensor.TYPE_LIGHT);
    text= (TextView) findViewById(R.id.text);
}
```

还需要在 Activity 的 onResume() 方法中，为传感器设置监听器。使用 registerListener()方法来注册一个 SensorEventListener 监听器，其中需要一个参数表示光线发生变化时获取数据并且通知用户的频率，这个参数就是传感器的采样频率，其中包括以下几个选择。

(1) SENSOR_DELAY_FASTEST：以最快的速度获得传感器数据($0\mu s$)；

(2) SENSOR_DELAY_GAME：适合于在游戏中获得传感器数据($20\,000\mu s$)；

(3) SENSOR_DELAY_UI：适合于在 UI 控件中获得传感器数据($60\,000\mu s$)；

(4) SENSOR_DELAY_NORMAL：选择默认的更新频率($200\,000\mu s$)。

上面 4 种类型获得传感器数据的速度依次递减。从理论上说，获得传感器数据的速度越快，在短时间内产生的事件就越多，需要消耗的系统资源越大。而有些传感器确实需要尽可能快的数据采集频率，特别是旋转矢量传感器，因此需要根据实际情况选择适当的速度获得传感器的数据。具体设置的代码如下：

```java
@Override
protected void onResume() {
    mgr.registerListener(this, light, SensorManager.SENSOR_DELAY_NORMAL);
    super.onResume();
}
```

在 onResume()和 onPause()回调方法中分别注册和注销传感器事件监听器是重要的一步。这是一种推荐的做法，即不需要的时候，不接收传感器数据，特别是在 Activity 暂停时。否则，有些传感器会大量地消耗电量，影响电池的寿命。对于 Android 系统的一些旧版本(例如 Android 2.1)，当屏幕关闭时传感器也会跟着关闭，但是比较新的版本

（例如 Android 4.1）则不会自动关闭。

如果应用程序需要在待机的状态下（屏幕关闭）也可以接收到传感器数据，就需要考虑保持传感器的不断更新，例如使用加速度计来检测设备的移动，代码如下：

```
@Override
protected void onPause() {
    mgr.unregisterListener(this, light);
    super.onPause();
}
```

可以为 Activity 实现 SensorEventListener 监听器接口，所以需要覆写两个回调方法，分别为 onAccuracyChanged() 和 onSensorChanged()。当传感器的准确性更改时，将调用 onAccuracyChanged(int sensor, int accuracy) 方法，参数 sensor 表示传感器对象、accuracy 表示该传感器新的准确度。其中精度包括高、低、中、不可靠，其值为 3、2、1 和 0，其常量分别为 SENSOR_STATUS_ACCURACY_HIGH、SENSOR_STATUS_ACCURACY_MEDIUM、SENSOR_STATUS_ACCURACY_LOW 和 SENSOR_STATUS_UNRELIABLE。代码如下：

```
@Override
public void onAccuracyChanged(Sensor sensor, int accuracy) {
    msg.insert(0, sensor.getName()
        +" accuracychanged: "
        +accuracy
        + (accuracy==1 ? " (LOW)" : (accuracy==2 ? " (MED)"
            : " (HIGH)"))+"\n");
    text.setText(msg);
    text.invalidate();
}
```

不可靠的精确度不代表传感器已经坏了，有可能是需要校准了。当光线发生变化时就会执行 onSensorChanged() 方法，可以得到传感器事件，其中包含了传感器获得的新值。onSensorChanged() 方法的参数是一个类型为 SensorEvent 的传感器事件对象，这个对象中的 values 变量非常重要，该变量的类型是 float[]。但该变量最多只有三个元素，而且根据传感器的不同，values 变量中元素所代表的含义也不同。对于光线传感器来说，只有 values 数组的第一个值有意义，它代表了传感器检测到的光线 SI lux（照明度单位）值。我们使用了 StringBuilder 记录光线变化的数据，并且通过 Textview 显示在界面上，代码如下：

```
@Override
public void onSensorChanged(SensorEvent event) {
    msg.insert(0, "Got a sensor event: "+event.values[0]
        +" SI lux units\n");
    text.setText(msg);
    text.invalidate();
```

}

需要在真机上运行上面的代码,因为虚拟机上没有光线传感器。光线传感器一般在手机上端,如果仔细观察的话,就会在上端屏幕后面看到一个小点,这就是光线传感器。如果用手指将此处覆盖,就会发现界面上显示的数值发生了变化。

或许读者会发现,没有直接的方法来查询传感器的值,只能通过注册监听器的方式从传感器获得最新的值。这意味着不能保证在一个确定的时间内获得新的数据。至少这个回调方法应该是异步的,目的是不会因为等待接收数据而阻塞 UI 线程。当然,可以通过本地代码和 Android 的 JNI 功能,直接访问传感器数据,但是这样做相对比较复杂,不在本书的讨论范围内。

7.3 传感器坐标系统

现在我们已经知道从传感器上获取数据的方法了,但是数据的表达方式是与具体的传感器相关的。通常,移动和位置传感器都使用了标准的三轴坐标系统来表示数据的值。这种坐标系统是相对设备屏幕的,默认情况下的手机位置如图 7-1 所示。

此时,x 轴表示从左到右的水平方向,y 轴表示自下而上的垂直方向,z 轴表示相对屏幕表面由内而外的方向。在这一坐标系中,屏幕背后的坐标用 z 轴的负值表示。会用到该坐标系的传感器包括加速度计、重力传感器、陀螺仪、线性加速度计和地磁传感器。要理解这个坐标系,最重要的一点就是,屏幕方向变化时坐标轴并不移动,也就是说设备移动时传感器的坐标系永不改变。这与 OpenGL 坐标系类似。理解坐标系的另一个要点——应用不得假定设备的初始(默认)方向是竖直

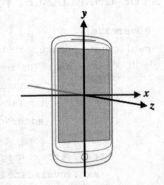

图 7-1 默认情况下的手机位置

的,很多桌面设备的初始方向是横向放置的。传感器的坐标系总是以设备的初始方向为基准。如果应用需要把传感器数据与屏幕显示关联,那么要用 getRotation() 方法来确定屏幕的转动方向,然后用 remapCoordinateSystem() 方法把传感器坐标映射为屏幕坐标。即使 AndroidManifest.xml 文件已经指定为仅支持纵向显示,仍需这么做。

7.4 详解各种传感器

7.4.1 加速度计

在移动设备中集成的加速度计可能是最有趣的传感器了,因为很多软件传感器都使用了加速度计。利用这个传感器,可以为应用程序提供设备的移动状态和位置状态数据。应用程序使用这些数据可以做各种有趣的事情,从玩游戏到增强现实,使移动设备变得更加智能。例如,移动设备经常用到的屏幕切换功能,就是当用户在竖屏和横屏之

间切换时,Android 系统获得加速度计传递的位置数据,然后切换用户界面。

加速度计的测量单位为米每秒平方(m/s^2)。地球重力加速度为 $9.81m/s^2$,方向向下并指向地球的中心。对于加速度计的坐标原点来说,重力加速度的测量值为 -9.81。如果设备完全处于静止状态(不移动),并且是在一个完全平坦的表面上,x 和 y 轴的读数将是 0,而 z 轴的读数是 9.81。实际上,由于加速度计灵敏度和准确度的问题,实测值可能会有误差,但它们将会接近。当设备是静止时,重力是唯一作用在设备上的力,如果设备是完全平放的,x 轴和 y 轴的效果是零,在 z 轴方向,加速度计测量的力量在设备上减去重力。因此,0 减 -9.81 是 9.81,也就是 z 值。

如果将设备平放,然后将它直线上升,则 z 值将首先增加,因为我们增加了 z 轴向上方向的力量。如果设备自由落体(最好不要这样),将向地面加速,所以加速度计将读取为 0。如下代码对加速度计传感器进行了测试。

```java
public class AccelerometerActivity extends Activity implements
        SensorEventListener {
    private SensorManager mgr;
    private Sensor accelerometer;
    private TextView text;
    private int mRotation;

    @Override
    public void onCreate(Bundle savedInstanceState) {
        super.onCreate(savedInstanceState);
        setContentView(R.layout.c08_sensor_accelerometer);
        mgr=(SensorManager) this.getSystemService(SENSOR_SERVICE);
        accelerometer=mgr.getDefaultSensor(Sensor.TYPE_ACCELEROMETER);
        text=(TextView) findViewById(R.id.text);
        WindowManager window=(WindowManager) this
                .getSystemService(WINDOW_SERVICE);
        int apiLevel=Integer.parseInt(Build.VERSION.SDK);
        if (apiLevel<8) {
            mRotation=window.getDefaultDisplay().getOrientation();
        } else {
            mRotation=window.getDefaultDisplay().getRotation();
        }
    }

    @Override
    protected void onResume() {
        mgr.registerListener(this, accelerometer,
            SensorManager.SENSOR_DELAY_UI);
        super.onResume();
    }
```

```
@Override
protected void onPause() {
    mgr.unregisterListener(this, accelerometer);
    super.onPause();
}

public void onAccuracyChanged(Sensor sensor, int accuracy) {
    //ignore
}

public void onSensorChanged(SensorEvent event) {
    String msg=String.format(
        "X: %8.4f\nY: %8.4f\nZ: %8.4f\nRotation: %d", event.values[0],
        event.values[1], event.values[2], mRotation);
    text.setText(msg);
    text.invalidate();
}
```

上面的代码在 Android 设备上运行后,拿起设备,旋转变成垂直的竖屏模式,y 值近似 9.81,x 值和 z 值近似 0;当设备旋转为横屏模式,并继续保持垂直时,y 值和 z 值近似 0,而 x 值近似 9.81。上面程序的运行效果如图 7-2 所示。

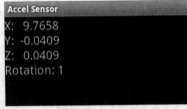

图 7-2 加速度计测试程序运行效果

加速度计使用标准的传感器坐标系。将移动设备平放在桌子上时:

(1) 如果推动设备的左边,向右侧移动设备,x 的加速度值是正值;

(2) 如果推动设备的底部,离开我们时,y 加速度值是正值;

(3) 如果朝上以 Am/s^2 加速度移动设备,z 轴的加速度值等于 $A+9.81$,对应于该设备的加速度($+Am/s^2$)减去重力加速度($-9.81m/s^2$);

(4) 一个静止在桌面上的设备加速度为 9.81,这对应于 $0m/s^2$ 减去重力加速度 $-9.81m/s^2$。

在一般情况下,如果我们正在监视设备运动,加速度计是一个很好的传感器,几乎所有的 Android 系统的手机和平板电脑都有加速度计。

7.4.2 重力传感器

Android 2.3 推出了基于加速度计的重力传感器,它不是一个真正的硬件传感器,而是一个虚拟传感器。重力传感器提供了三维矢量,指示重力的方向和大小。以下代码演示了如何获得默认的重力感应器。

```
private SensorManager mSensorManager;
private Sensor mSensor;
...
```

```
mSensorManager= (SensorManager) getSystemService(Context.SENSOR_SERVICE);
mSensor=mSensorManager.getDefaultSensor(Sensor.TYPE_GRAVITY);
```

重力传感器采用与加速度计相同的单位(m/s^2)和坐标系。当一个设备处于静止状态时,重力传感器的输出值应该与加速度计的输出值相同。

7.4.3 陀螺仪

陀螺仪又叫角速度传感器。不同于加速度计,陀螺仪测量的物理量是偏转、倾斜时的转动角速度。在手机上,仅用加速度计没办法测量或重构出完整的 3D 动作,加速度计只能检测轴向的线性动作,而测量不到转动的动作;但陀螺仪则可以对转动、偏转的动作进行测量,这样就可以精确分析从而判断出移动设备的实际动作。陀螺仪测量单位为 rad/s,也是依据 x 轴、y 轴和 z 轴的旋转和速率得到的。以下代码演示了如何得到一个实例的默认陀螺仪。

```
private SensorManager mSensorManager;
private Sensor mSensor;
...
mSensorManager= (SensorManager) getSystemService(Context.SENSOR_SERVICE);
mSensor=mSensorManager.getDefaultSensor(Sensor.TYPE_GYROSCOPE);
```

陀螺仪的坐标系统与加速度计是相同的,当逆时针方向旋转时为正值。

7.4.4 线性加速度

线性加速度计提供移动设备每个轴的三维矢量加速度,而且不计重力。以下代码演示了如何获得线性加速度计的默认实例:

```
private SensorManager mSensorManager;
private Sensor mSensor;
...
mSensorManager= (SensorManager) getSystemService(Context.SENSOR_SERVICE);
mSensor=mSensorManager.getDefaultSensor(Sensor.TYPE_LINEAR_ACCELERATION);
```

从概念上讲,这种传感器提供的加速度数据根据下列关系式计算:

$$线性加速度 = 加速度 - 重力加速度$$

这个传感器的典型应用是获取去除了重力干扰的加速度数据。例如,可以用这个传感器来获取汽车加速度。线性加速度传感器总是会有些偏差,需要把这个偏差值抵消掉。最简单的消除方式就是在应用中增加一个校准的环节。在校准过程中,可以要求用户先把设备放在桌子上,再来读取三个坐标轴的偏差值。然后就可以从传感器的读数中减去这个偏差值,以获取真实的线性加速度。这个传感器的坐标系与加速度传感器使用的相同,单位也是 m/s^2。

7.4.5 方向传感器

方向传感器用于监测设备相对于地球的方位(地球磁场)。方向传感器的数据来自设备的地磁传感器和加速度计。利用这两种硬件传感器,方向传感器提供了以下三个方向的数据。

1. 侧倾度(围绕 z 轴的旋转角)

这是指设备 y 轴与地磁北极间的夹角。例如,如果设备的 y 轴指向地磁北极则该值为 0,如果 y 轴指向南方则该值为 180。同理,y 轴指向东方则该值为 90,而指向西方则该值为 270。

2. 俯仰度(围绕 x 轴的旋转角)

当 z 轴的正值部分朝向 y 轴的正值部分旋转时,该值为正。当 z 轴的正值部分朝向 y 轴的负值部分旋转时,该值为负。取值范围为 $-180°\sim180°$。

3. 翻滚度(围绕 y 轴的旋转角)

当 z 轴的正值部分朝向 x 轴的正值部分旋转时,该值为正;当 z 轴的正值部分朝向 x 轴的负值部分旋转时,该值为负。取值范围为 $-90°\sim90°$。翻滚度也是以顺时针方向为正(从数学上讲,应该是逆时针方向为正)。

方向传感器的数据是对加速度和地磁传感器的原始数据进行处理之后再报送出来的。因为处理工作比较繁重,方向传感器的精度和准确度会有所降低(只有在翻滚度为 0 时此传感器的数据才是可靠的)。因此,方向传感器自 Android 2.2(API level 8)开始已经过时了。作为直接使用方向传感器原始数据的替代方案,建议结合 getRotationMatrix() 和 getOrientation() 方法来计算方向值。还可以用 remapCoordinateSystem() 方法来把方向值转换为应用程序自定义参照系的坐标。下面的代码示例显示了如何直接从方向传感器获得定位数据。

```java
public class SensorActivity extends Activity implements SensorEventListener {

    private SensorManager mSensorManager;
    private Sensor mOrientation;

    @Override
    public void onCreate(Bundle savedInstanceState) {
        super.onCreate(savedInstanceState);
        setContentView(R.layout.main);
        mSensorManager= (SensorManager)
            getSystemService(Context.SENSOR_SERVICE);
        mOrientation=mSensorManager
            .getDefaultSensor(Sensor.TYPE_ORIENTATION);
    }

    @Override
```

```java
public void onAccuracyChanged(Sensor sensor, int accuracy) {
    //Do something here if sensor accuracy changes
    //You must implement this callback in your code
}

@Override
protected void onResume() {
    super.onResume();
    mSensorManager.registerListener(this, mOrientation,
            SensorManager.SENSOR_DELAY_NORMAL);
}

@Override
protected void onPause() {
    super.onPause();
    mSensorManager.unregisterListener(this);
}

@Override
public void onSensorChanged(SensorEvent event) {
    float azimuth_angle=event.values[0];
    float pitch_angle=event.values[1];
    float roll_angle=event.values[2];
    //Do something with these orientation angles.
}
}
```

7.4.6 地磁场传感器

地磁场传感器可让我们监视地球的磁场变化。以下代码演示了如何获得默认的地磁场传感器的实例。

```java
private SensorManager mSensorManager;
private Sensor mSensor;
...
mSensorManager= (SensorManager) getSystemService(Context.SENSOR_SERVICE);
mSensor=mSensorManager.getDefaultSensor(Sensor.TYPE_MAGNETIC_FIELD);
```

这种传感器提供原始的三个坐标轴的磁场强度数据(μT)。通常情况下,不需要直接使用这种传感器。相反,可以使用旋转矢量传感器,以确定设备的旋转运动。可以利用加速度计和地磁场传感器配合 getRotationMatrix()方法获得旋转矩阵和倾斜矩阵。然后,可以使用这些矩阵的 getOrientation()和 getInclination()方法,获得方位角和磁倾角的数据。

7.4.7 距离传感器

距离传感器使我们能检测设备距离某物体的远近程度,通常用于确定用户头部与手持设备屏幕表面的距离(例如用户拨打或接听电话时)。大部分距离传感器返回的是绝对距离,单位是 cm,但某些传感器只能返回远近程度值,就是代表远近程度的二进制数值。这种情况下,传感器通常把最大量程表示为"远"、小于量程的值则为"近"。"远"值典型为>5cm,但这因传感器而异。可以用 getMaximumRange()方法来确定传感器的最大量程。以下代码演示了如何获得一个实例的默认距离传感器。

```
private SensorManager mSensorManager;
private Sensor mSensor;
...
mSensorManager= (SensorManager) getSystemService(Context.SENSOR_SERVICE);
mSensor=mSensorManager.getDefaultSensor(Sensor.TYPE_PROXIMITY);
```

距离传感器通常用于确定手持通话器设备离一个人的头部的远近。手机上的距离传感器一般在前置摄像头附近,也就是正面、屏幕的上方。在接电话的时候距离传感器会起作用,当脸部靠近屏幕,屏幕灯会熄灭,并自动锁屏,可以防止脸部误操作;当脸部离开,屏幕灯会自动开启,并且自动解锁。以下代码演示了如何使用脸部传感器。

```
public class SensorActivity extends Activity implements SensorEventListener {
    private SensorManager mSensorManager;
    private Sensor mProximity;

    @Override
    public final void onCreate(Bundle savedInstanceState) {
        super.onCreate(savedInstanceState);
        setContentView(R.layout.main);

        //Get an instance of the sensor service, and use that to get
        //an instance of a particular sensor
        mSensorManager= (SensorManager) getSystemService(Context
            .SENSOR_SERVICE);
        mProximity=mSensorManager.getDefaultSensor(Sensor.TYPE_PROXIMITY);
    }

    @Override
    public final void onAccuracyChanged(Sensor sensor, int accuracy) {
        //Do something here if sensor accuracy changes
    }

    @Override
    public final void onSensorChanged(SensorEvent event) {
```

```
        float distance=event.values[0];
        //Do something with this sensor data
    }

    @Override
    protected void onResume() {
        //Register a listener for the sensor
        super.onResume();
        mSensorManager.registerListener(this, mProximity,
            SensorManager.SENSOR_DELAY_NORMAL);
    }

    @Override
    protected void onPause() {
        //Be sure to unregister the sensor when the activity pauses
        super.onPause();
        mSensorManager.unregisterListener(this);
    }
}
```

一些距离传感器返回的是表示"近"或"远"的二进制值,在这种情况下,该传感器通常报告其最大范围值在远的状态和在不久的状态下一个更低的值。可以通过使用getMaximumRange()方法,来决定一个传感器的最大范围。

习 题 7

1. 哪个传感器可以用于制作微信里的"摇一摇"功能?请简要说明其实现方法。
2. Android 平台提供了哪几种类型的传感器?
3. 如何设置传感器的监听器?

第 8 章

Android 服务简介

Service(服务)是 Android 四大组件中与 Activity 很相似的组件,它们都代表可执行的程序,都是从 Context 派生出来的。Service 与 Activity 的区别在于:Service 一直在后台运行,它没有用户界面。一旦 Service 被启动起来之后,它就与 Activity 一样。

8.1 Service 的创建及配置

与 Activity 创建和配置一样,Service 也需要先创建一个继承 Service 的子类,然后在 AndroidManifest.xml 文件中配置该 Service。如图 8-1 所示,在创建 Service 时要指定继承 Android 的 Service 服务类。

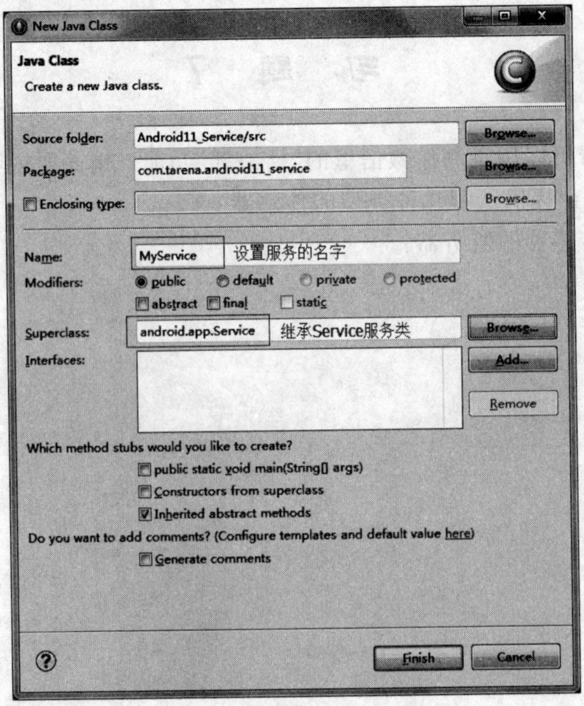

图 8-1 创建 Service 服务类

创建 Service 之后，一般要对 AndroidManifest.xml 文件进行配置，如：

```
<service android:name="MyService">
</service>
```

8.2 Service 的分类及生命周期

在应用中往往有一些文件需要下载，或者是一些通信信息需要等待其返回，这些动作都是极为耗时的操作，这时就要考虑执行该动作的组件能够长时间运行（包括应用隐藏在手机后台等），而这些动作最适合的组件就是 Service 服务。因为它的特点就是能够长时间运行在应用程序后台。

8.2.1 Service 分类

Service 的生命周期从它被创建开始，到它被销毁为止，可以分为两类：本地服务和绑定本地服务（绑定服务）。

1. 本地服务

被开启的 Service 通过其他组件调用 startService() 被创建。这种 Service 可以无限地运行下去，必须调用自身的 stopSelf() 方法或者其他组件调用 stopService() 方法来停止它。当 Service 被停止时，系统会销毁它。这种 Service 同时需要具备自管理能力，且不需要通过函数的调用向外部提供数据或功能。

2. 绑定本地服务

被绑定的 Service 是其他组件调用 bindService() 来创建的。客户可以通过一个 IBinder 接口和 Service 进行通信，客户也可以通过 unbindService() 方法来关闭这个服务，同时解除绑定。

需要注意，如果绑定过程中 Service 没有被启动，Context.bindService() 会自动启动 Service。

上述两种 Service 并不是完全独立的，在某种情况下可以混合使用。例如一个音乐播放器，在后台工作的 Service 通过 startService() 启动某个特定音乐播放，但在音乐播放当中用户需要暂停音乐播放，则需要通过 bindService() 获取服务连接和 Service 对象，进而通过调用 Service 对象中的函数暂停音乐播放过程，并保存相关信息。在这种情况下，如果调用 stopService() 并不能停止 Service，需要在所有的服务连接关闭后，Service 才能够真正停止。

8.2.2 Service 生命周期

和 Activity 一样，Service 也有一系列的生命周期回调方法，可以实现它们来监测 Service 状态的变化，并且在适当的时候执行适当的处理逻辑。

下面的代码片段中展示了 Service 每一个生命周期的方法。

```java
public class MyService extends Service {
    @Override
    public IBinder onBind(Intent arg0) {
        returnnull;
    }
    @Override
    public void onCreate() {
        super.onCreate();
    }
    @Override
    public void onDestroy() {
        super.onDestroy();
    }
    @Override
    public int onStartCommand(Intent intent, int flags, int startId) {
        returnsuper.onStartCommand(intent, flags, startId);
    }
    @Override
    public boolean onUnbind(Intent intent) {
        return super.onUnbind(intent);
    }
}
```

从上述代码中可以知道,Service 的生命周期包含了如下 5 个方法。

(1) onCreate():Service 的生命周期开始方法,主要完成 Service 初始化方法。

(2) onStartCommand():活动生命周期开始,但没有对应的 Stop 方法。

(3) onBind():绑定服务开始方法。

(4) onUnbind():解除绑定服务结束方法。

(5) onDestroy():服务结束方法。

Service 生命周期方法之间的关系,如图 8-2 所示。

图 8-2 说明了 Service 的典型回调方法,尽管图中将开启的 Service 和绑定的 Service 分开,但是需要记住,任何 Service 都潜在地允许绑定。所以,一个被开启的 Service 仍然可能被绑定。

实现这些方法,可以看到两层嵌套的 Service 生命周期。

1. 整体生命周期

Service 的整体生命周期是从 onCreate() 被调用开始,到 onDestroy() 方法返回为止。和 Activity 一样,Service 在 onCreate() 中进行它的初始化工作,在 onDestroy() 中释放残留的资源。

onCreate() 和 onDestroy() 会被所有的 Service 调用,不论 Service 是通过

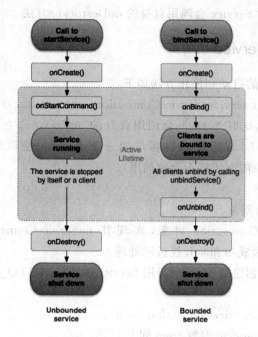

图 8-2　Service 的生命周期

startService()还是 bindService()建立。

2．活动的生命周期

　　Service 积极活动的生命时间（Active Lifetime）是从 onStartCommand()或 onBind()被调用开始，它们各自处理由 startService()或 bindService()方法传过来的 Intent 对象。

　　如果 Service 是被开启的，那么它的活动生命周期和整个生命周期一同结束；如果 Service 是被绑定的，它的活动生命周期是在 onUnbind()方法返回后结束。

　　需要注意，尽管一个被开启的 Service 是通过调用 stopSelf()或 stopService()来停止的，但没有一个回调方法与之对应，即没有 onStop()回调方法。所以，当调用了停止的方法，除非这个 Service 和客户组件绑定，否则系统将会直接销毁它，这时 onDestory()方法会被调用，并且是这个时候唯一会被调用的回调方法。

8.3　启动和停止 Service

8.3.1　本地 Service

　　本地 Service 的启动和停止方法如下。

　　（1）startService(Intent intent)；

　　启动 Service，这时 Service 会调用自身的 onCreate()方法（该 Service 未创建时），接着调用 onStartCommand()方法。

　　（2）stopService(Intent intent)；

停止 Service,这时 Service 会调用自身的 onDestory()方法。

8.3.2 绑定本地 Service

绑定本地 Service 的启动和停止方法如下。

bindService(Intent service,ServiceConnection conn,int flags);

绑定一个 Service,这时 Service 会调用自身的 onCreate()方法(该 Service 未创建时),接着调用 onBind()方法返回给客户端一个 IBinder 接口对象(如果返回 null,ServiceConnection 对象的方法将不会被调用)。

方法中的参数如下。

① service：Intent 对象。

② conn：ServiceConnection 对象,实现其 onServiceConnected()和 onServiceDisconnected(),在连接成功和断开连接时处理。

③ flags：Service 创建的方式,一般用 Service.BIND_AUTO_CREATE 表示绑定时自动创建。

④ unbindService(ServiceConnection conn);

解除绑定的一个 Service,参数 conn 同上。

8.3.3 Service 案例

案例界面有 5 个按钮分别是 startService 按钮、stopService 按钮、bindService 按钮、unbindServec 按钮和 callService 按钮,其功能分别是启动本地 Service、停止本地 Service、绑定 Service、解除绑定的 Service 和调用 Service 中的方法。

案例中包括一个布局文件(activity_main.xml),一个 Activity 组件类(MainActivity 类)和一个 Service 组件类(MyService)。案例代码如下。

1. 布局文件 activity_main.xml

```xml
<LinearLayout xmlns:android="http://schemas.android.com/apk/res/android"
    xmlns:tools="http://schemas.android.com/tools"
    android:id="@+id/LinearLayout1"
    android:layout_width="match_parent"
    android:layout_height="match_parent"
    android:orientation="vertical"
    android:paddingBottom="@dimen/activity_vertical_margin"
    android:paddingLeft="@dimen/activity_horizontal_margin"
    android:paddingRight="@dimen/activity_horizontal_margin"
    android:paddingTop="@dimen/activity_vertical_margin"
    tools:context=".MainActivity">

    <Button
        android:id="@+id/btn_startService"
```

```xml
        android:layout_width="match_parent"
        android:layout_height="wrap_content"
        android:text="startService" />

    <Button
        android:id="@+id/btn_stopService"
        android:layout_width="match_parent"
        android:layout_height="wrap_content"
        android:text="stopService" />

    <Button
        android:id="@+id/btn_bindService"
        android:layout_width="match_parent"
        android:layout_height="wrap_content"
        android:text="bindService" />

    <Button
        android:id="@+id/btn_callService"
        android:layout_width="match_parent"
        android:layout_height="wrap_content"
        android:text="callService" />

    <Button
        android:id="@+id/btn_unbindService"
        android:layout_width="match_parent"
        android:layout_height="wrap_content"
        android:text="unbindService" />

</LinearLayout>
```

2. Activity 组件 MainActivity 类

```java
package com.example.android_demo8_1;

import com.example.android_demo8_1.MyService.MyBind;
import android.app.Activity;
import android.content.ComponentName;
import android.content.Context;
import android.content.Intent;
import android.content.ServiceConnection;
import android.os.Bundle;
import android.os.IBinder;
import android.util.Log;
import android.view.View;
```

```java
import android.view.View.OnClickListener;
import android.widget.Button;

/**
 *服务类
 *
 */
public class MainActivity extends Activity implements OnClickListener{
    //按钮控件对象
    Button startService, stopService, bindService, unbindService, callService;
    //Log TAG
    private static final String TAG="MainActivity";
    MyBind mBind;

    @Override
    protected void onCreate(Bundle savedInstanceState) {
        super.onCreate(savedInstanceState);
        //设置布局
        setContentView(R.layout.activity_main);
        //获得控件对象
        startService=(Button) this.findViewById(R.id.btn_startService);
        stopService=(Button) this.findViewById(R.id.btn_stopService);
        bindService=(Button) this.findViewById(R.id.btn_bindService);
        unbindService=(Button) this.findViewById(R.id.btn_unbindService);
        callService=(Button) this.findViewById(R.id.btn_callService);
        //添加监听器方法
        startService.setOnClickListener(this);
        stopService.setOnClickListener(this);
        bindService.setOnClickListener(this);
        unbindService.setOnClickListener(this);
        callService.setOnClickListener(this);
    }

    @Override
    public void onClick(View view) {
        Intent intent=new Intent();
        intent.setClass(MainActivity.this, MyService.class);
        switch (view.getId()) {
        case R.id.btn_startService:
            //开启服务
            Bundle mbundle=new Bundle();
            mbundle.putString("hello", "你好吗,myService");
            intent.putExtras(mbundle);
            startService(intent);
```

```java
            break;
        case R.id.btn_stopService:
            //关闭服务
            Bundle umbundle=new Bundle();
            umbundle.putString("hello", "再见,myService");
            intent.putExtras(umbundle);
            stopService(intent);
            break;
        case R.id.btn_bindService:
            //开启绑定服务
            Bundle bmbundle=new Bundle();
            bmbundle.putString("hello", "你好吗,myService");
            intent.putExtras(bmbundle);
            bindService(intent, conn, Context.BIND_AUTO_CREATE);
            break;
        case R.id.btn_unbindService:
            //解除绑定服务
            unbindService(conn);
            break;
        case R.id.btn_callService:
            //调用服务提供的方法
            String str=mBind.ActivityCallService();
            Log.i(TAG, str);
            break;
    }
}

//实现 ServiceConnection 接口对象
ServiceConnection conn=new ServiceConnection() {

    @Override
    public void onServiceDisconnected(ComponentName arg0) {
        //
    }

    @Override
    public void onServiceConnected(ComponentName arg0, IBinder arg1) {
        Log.e(TAG, "连接成功");
        //Service 连接建立成功后,提供给客户端与 Service 交互的对象
        mBind= (MyBind) arg1;
    }
};
}
```

3. Service 组件 MyService 类

```java
package com.example.android_demo8_1;

import android.app.Service;
import android.content.Intent;
import android.os.Binder;
import android.os.Bundle;
import android.os.IBinder;
import android.util.Log;
/**
 * 服务类
 *
 */
public class MyService extends Service {

    private static final String TAG="MyService";
    public MyBind myBinder;

    public class MyBind extends Binder {
        //公开方法,提供给 Activity 调用
        public String ActivityCallService() {
            Log.i(TAG, "Activity Call Service");
            return "Activity Call Service success";
        }
    }

    @Override
    public IBinder onBind(Intent arg0) {
        Log.i(TAG, "onBind");
        Bundle bundle=(Bundle)arg0.getExtras();
        String keyvalue=bundle.getString("hello");
        Log.i(TAG, keyvalue);
        return myBinder;
    }
    @Override
    public void onCreate() {
        Log.i(TAG, "onCreate");
        myBinder=new MyBind();
        super.onCreate();
    }
    @Override
    public void onDestroy() {
        Log.i(TAG, "onDestroy");
```

```
        super.onDestroy();
    }
    @Override
    public int onStartCommand(Intent intent, int flags, int startId) {
        Log.i(TAG, "onStartCommand");
        Bundle bundle= (Bundle)intent.getExtras();
        String keyvalue=bundle.getString("hello");
        Log.i(TAG, keyvalue);
        return super.onStartCommand(intent, flags, startId);
    }
    @Override
    public boolean onUnbind(Intent intent) {
        Log.i(TAG, "onUnbind");
        return super.onUnbind(intent);
    }
}
```

案例运行效果如图 8-3 所示。在案例运行之后,从上到下依次单击界面中的按钮,LogCat 显示如图 8-4 所示。

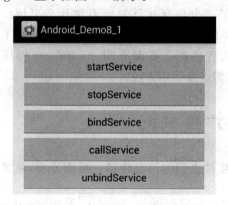

图 8-3　Service 案例运行效果　　　　图 8-4　Service 案例运行 LogCat 显示

习　题　8

1. Service 与 Activity 有什么区别?
2. 如何开启和停止 Service?
3. 简要说明 Service 常用的生命周期回调方法?

第 9 章 Android 广播简介

BroadcastReceiver(广播接收者)为 Android 的四大组件之一,主要用来接收广播 Intent。在 Android 中,广播是一种广泛运用在应用程序之间传输信息的机制。而 BroadcastReceiver 是对发送出来的广播进行过滤接收并响应的一类组件。例如,我买了一张从北京到哈尔滨的 T41 次火车票,由于我去火车站过早,于是就去候车室周围的图书馆看书了,而就在我看书的时候,听到火车站广播中说:"前往哈尔滨方向的 T41 次列车开始检票了",于是我就赶往候车室检票上车。通过刚才的例子可以了解到生活中的广播,其实就是通知我们一些信息,当我听到了这个信息我就知道要做什么事情。而 Android 的广播和生活中的广播工作原理是一样的,就是告诉程序它应该干什么,只不过生活中是发送给人的,程序中发送的广播是给程序的。下面介绍 Android 中提供的 BroadcastReceiver(广播接收者),让读者了解其特点及使用方法。

9.1 Android 广播机制

Android 中使用广播这种异步机制来处理组件之间的消息传递。所谓异步,就是广播的发送方和接收方不需要相互等待。

了解 Android 中广播的特点,那广播使用的是什么机制呢?其实在 Android 中广播采用的是订阅-发送机制,属于设计模式中的观察者模式(Observer)。在广播的底层实现中,系统为广播发送方维护了一个目标列表,每次要发送广播时,发送方就会遍历这个列表,对其中的每一个目标发送广播。想要接收某个广播的所有接收方,都要事先在该广播对应的列表中完成注册,才能接收到这个广播。

9.2 收发广播

9.2.1 发送广播

在应用中如果需要发送广播,需要定义一个 Intent 对象,用于封装要发送的消息,并指定 Intent 中的 Action 属性用于匹配,然后再使用 Context.sendBroadCast()方法,将 Intent 对象发送出去,参考代码片段如下:

```
Intent intent=new Intent();
intent.setAction("Action");
intent.putExtra("username","张三");
this.sendBroadcast(intent);
```

9.2.2 接收广播

接收广播的时候,使用 BroadcastReceiver 类,需要实现其内部的 onReceive()方法,并在该方法中做具体的操作。参考代码片段如下:

```
private BroadcastReceiver receiver=new BroadcastReceiver(){
    @Override
    public void onReceive(Context context, Intent intent){
        String action=intent.getAction();
        if("Action".equals(action)){
            Log.i("info","Action:"+action);
        }
    }
}
```

上述代码中,可以通过 Intent 对象获得传递到接收者中的数据,如 Action、Data 等数据。

9.2.3 BroadcastReceiver(广播接收者)注册分类

在 Android 中,把广播的注册分为两类,一类是静态注册广播,另一类是动态注册广播。静态注册广播就是预先注册好放在那里等待使用,而动态注册广播就是什么时候需要什么时候注册使用。

1. 静态注册广播

常驻型广播也就是静态注册广播,当应用程序关闭了,如果有广播信息过来,广播接收器同样能接收到,它的注册方式就是在应用程序的 AndroidManifest.xml 中进行注册,这种注册方式通常又被称作静态注册。需要使用 BroadcastReceiver 时,首先要在对应的包中创建继承 BroadcastReceiver 类,并将其通过 <receiver> 标签注册到 AndroidManifest.xml 文件中。参考代码片段如下:

```
<receiver android:name="MyBroadcastReceiver">
<intent-filter>
<action android:name="com.example.android_demo8_1_broadcastreceiver01" />
</intent-filter>
</receiver>
```

上述代码中通过<intent-filter>标签的 action 动作找到对应的广播接收者。

2. 动态注册广播

非常驻型广播也就是动态注册广播,当应用程序结束了,广播自然就没有了,例如在 Activity 中的 onCreate 或者 onResume 中注册广播接收者,在 onDestory 中注销广播接收者。这样广播接收者就是一个非常驻型的广播了,这种注册方式也叫动态注册。动态注册需要在代码中设置一个 IntentFilter 对象,然后在需要注册的地方调用 Context.registerReceiver()方法,当取消时就调用 Context.unregisterReceiver()方法。

9.2.4 静态注册广播案例

案例界面有一个文本框和一个按钮,单击按钮后会将文本框内容通过广播发送出去,程序接收到广播后会弹出 Toast 提示信息,显示输入框内容。

案例中包括一个布局文件(activity_main.xml),一个 Activity 组件类(MainActivity 类),一个 BroadcastReceiver 组件类(MyBroadcastReceiver),用于接收和处理广播。

案例程序需要在 AndroidManifest.xml 文件中注册广播接收者,如下:

```xml
<receiver android:name="MyBroadcastReceiver">
<intent-filter>
<action android:name="com.example.android_demo8_1_broadcastreceiver01" />
</intent-filter>
</receiver>
```

1. 布局文件 activity_main.xml 代码

```xml
<LinearLayout xmlns:android="http://schemas.android.com/apk/res/android"
    xmlns:tools="http://schemas.android.com/tools"
    android:id="@+id/LinearLayout1"
    android:layout_width="match_parent"
    android:layout_height="match_parent"
    android:orientation="vertical"
    android:paddingBottom="@dimen/activity_vertical_margin"
    android:paddingLeft="@dimen/activity_horizontal_margin"
    android:paddingRight="@dimen/activity_horizontal_margin"
    android:paddingTop="@dimen/activity_vertical_margin"
    tools:context=".MainActivity">

    <EditText
        android:id="@+id/et_content"
        android:layout_width="match_parent"
        android:layout_height="wrap_content"
        android:ems="10">

        <requestFocus />
    </EditText>
```

```xml
<Button
    android:id="@+id/btn_send_message"
    android:layout_width="match_parent"
    android:layout_height="wrap_content"
    android:layout_marginTop="10dp"
    android:text="发送广播"
    android:onClick="onClick"/>

</LinearLayout>
```

2. Activity 组件 MainActivity 类

```java
package com.example.android_demo9_1;

import android.app.Activity;
import android.content.Intent;
import android.os.Bundle;
import android.view.View;
import android.widget.EditText;

/**
 * 广播类
 */
public class MainActivity extends Activity {
    //编辑文本框对象
    private EditText contentEt;
    @Override
    protected void onCreate(Bundle savedInstanceState) {
        super.onCreate(savedInstanceState);
        //设置布局
        setContentView(R.layout.activity_main);
        //获得编辑文本框对象
        contentEt=(EditText) this.findViewById(R.id.et_content);
    }
    public void onClick(View view){
        //获得文本框数据
        String content=contentEt.getText().toString().trim();
        //发送广播
        Intent intent=new Intent();
        intent.setAction("com.example.android_demo8_1_broadcastreceiver01");
        intent.putExtra("content", content);
        sendBroadcast(intent);
    }
}
```

3. BroadcastReceiver 组件 MyBroadcastReceiver 类

```
package com.example.android_demo9_1;

import android.content.BroadcastReceiver;
import android.content.Context;
import android.content.Intent;
import android.widget.Toast;
/**
 * 广播接收者
 */
public class MyBroadcastReceiver extends BroadcastReceiver {

    @Override
    public void onReceive(Context context, Intent intent) {
        //获得发送数据
        String content=intent.getStringExtra("content");
        //提示
        Toast.makeText(context, "发送广播内容:"+content, Toast.LENGTH_LONG).show();
    }

}
```

案例运行效果如图 9-1 所示。

9.2.5 动态注册广播案例

案例 Activity 在启动时,动态注册一个系统的短信广播接收者,当系统收到短信后,案例程序会收到系统广播,之后会使用 Toast 提示短信内容。

案例中包括一个布局文件(activity_main.xml)和一个 Activity 组件类(MainActivity 类)。在 MainActivity 类中,创建了一个广播接收者成员,并复写其 onReceive 方法,在方法中通过传入的 Intent 对象获取短信内容,之后使用 Toast 显示短信内容。MainActivity 类还在 onCreate 方法中动态注册一个系统短信的广播接收者,在 onDestroy 方法中注销动态注册的系统短信广播接收者。

图 9-1 静态注册广播案例运行效果

1. 布局文件 activity_main.xml 代码

```
<RelativeLayout xmlns:android="http://schemas.android.com/apk/res/android"
    xmlns:tools="http://schemas.android.com/tools"
```

```xml
    android:layout_width="match_parent"
    android:layout_height="match_parent"
    android:paddingBottom="@dimen/activity_vertical_margin"
    android:paddingLeft="@dimen/activity_horizontal_margin"
    android:paddingRight="@dimen/activity_horizontal_margin"
    android:paddingTop="@dimen/activity_vertical_margin"
    tools:context=".MainActivity">

    <TextView
        android:layout_width="wrap_content"
        android:layout_height="wrap_content"
        android:text="通过广播拦截短信"
        android:textSize="30sp"/>

</RelativeLayout>
```

2. Activity 组件 MainActivity 类

```java
package com.example.android_demo9_2;

import android.app.Activity;
import android.content.BroadcastReceiver;
import android.content.Context;
import android.content.Intent;
import android.content.IntentFilter;
import android.os.Bundle;
import android.telephony.SmsMessage;
import android.util.Log;
import android.widget.Toast;

/**
 * 广播拦截短信
 */
public class MainActivity extends Activity {
    @Override
    protected void onCreate(Bundle savedInstanceState) {
        super.onCreate(savedInstanceState);
        //设置布局
        setContentView(R.layout.activity_main);
        //声明意图过滤器
        IntentFilter filter=new IntentFilter();
        //设置系统动作
        filter.addAction("android.provider.Telephony.SMS_RECEIVED");
        //设置优先级
```

```java
        filter.setPriority(1000);
        //注册
        registerReceiver(receiver, filter);
    }
    @Override
    protected void onDestroy() {
        //TODO Auto-generated method stub
        super.onDestroy();
        //取消注册
        unregisterReceiver(receiver);
    }
    /**
     * 声明广播接收者对象
     */
    BroadcastReceiver receiver=new BroadcastReceiver() {

        @Override
        public void onReceive(Context context, Intent intent) {
            //获得系统数据
            Bundle bundle=intent.getExtras();
            //判断数据是否为NULL
            if(bundle !=null){
                //获得短信原始数据
                Object[] objects=(Object[])bundle.get("pdus");
                //声明短信数组
                SmsMessage[] messages=new SmsMessage[objects.length];
                //循环包装数据
                for (int i=0; i<messages.length; i++) {
                    messages[i]=SmsMessage.createFromPdu((byte[])objects[i]);
                }
                //循环遍历短信数据
                for (SmsMessage smsMessage : messages) {
                    //获得电话号码
                    String body=smsMessage.getDisplayMessageBody();
                    //获得短信内容
                    String address=smsMessage.getDisplayOriginatingAddress();
                    Log.i("info", "Address-->"+address+",Body-->"+body);
                    Toast.makeText(context,
                        "Address-->"+address+",Body-->"+body,
                        Toast.LENGTH_LONG).show();
                }
            }
        }
    };
```

}

案例运行效果如图 9-2 所示。

图 9-2 动态注册广播案例运行效果

9.3 系统自带的广播

Android 系统提供了一些默认广播,它们会随着系统的一些变化而被发送出去,供应用程序接收及处理。一些常用的系统广播如表 9-1 所示。

表 9-1 常用系统广播

广播 Action	广播说明
ACTION_AIRPLANE_MODE_CHANGED	进入或退出飞行模式
ACTION_BATTERY_CHANGED	电池电量改变
ACTION_BATTERY_LOW	电量过低
ACTION_BATTERY_OKAY	在电量过低被发送之后电量又增加到 OK 状态
ACTION_BOOT_COMPLETED	系统启动完成,只广播一次
ACTION_CAMERA_BUTTON	照相机功能键被按下
ACTION_DEVICE_STORAGE__LOW	设备内存不足
ACTION_DEVICE_STORAGE_OK	设备内存不足状态消失
ACTION_INPUT_METHOD_CHANGED	输入法改变

续表

广播 Action	广播说明
ACTION_TIMEZONE_CHANGED	设备所处的时区改变
ACTION_MEDIA_EJECT	用户想要移除外部存储介质
ACTION_MEDIA_MOUNTED	外部媒介被加载
ACTION_MEDIA_REMOVED	外部设备被移除
ACTION_MEDIA_SHARED	外部媒介不能被加载,因为它们共享一个 USB 存储设备
ACTION_SCREEN_OFF	屏幕被关闭
ACTION_SCREEN_ON	屏幕被开启
ACTION_POWER_CONNECTED	外部充电电源连接上
ACTION_POWER_DISCONNECTED	外部充电设备被移除

9.4 广播分类

Android 中可供接收的广播有三种:正常广播、有序广播和黏滞广播。

9.4.1 正常广播

之前介绍的案例一直使用的就是正常广播,正常广播也称为无序广播,这种广播使用 Context.sendBroadcast()方法发送,是完全异步的。所有的接收器的执行顺序不确定,因此所有的接收器接收广播的顺序不确定。

9.4.2 有序广播

有序广播(Ordered Broadcast)调用 Context.sendOrderedBroadcast()方法发送,在同一时刻只能传送到一个接收器。多个接收器是依次执行的,每个处理完广播的接收器可以向下一个接收器传送一个结果,或者直接中止广播的传递,这样接下来的广播就接收不到广播了。

在接收器中可以通过设置 android:priority 属性来决定广播接收器的优先级,而这个属性需要设置在 AndroidMenifest.xml 文件中。如果接收器的优先级相同,接收顺序是不确定的。

9.4.3 黏滞广播

黏滞广播(Sticky Broadcast)调用 Context.sendStickyBroadcast()方法发送,它与正常广播很相似,只有一点区别。对于正常广播来说,需要先注册接收器,才能接收广播。如果一个正常广播发送之后,才注册广播接收器,则这个广播是不能被接收到的。但是对于黏滞广播来说,这种情况就可以接收到广播,当一个黏滞广播被发送之后,会自动保存下来,一旦有接收器注册这条广播,就会立即收到这条广播。注意:要是使用黏滞广播,需要在 AndoridMenifest.xml 文件中注册对应的权限,具体的代码片段如下:

```
<uses-permission android:name="android.permission.BROADCAST_STICKY"/>
```

习 题 9

1. 如何进行广播的发送与接收?
2. Android 中可供接收的广播有哪些?
3. 注册广播有几种方式,这些方式有何优缺点?
4. 说出三个 Android 系统提供的默认广播。

第 10 章

Android 的数据持久化

数据持久化就是把要操作的数据永久地保存起来,在需要使用的时候再把数据读取到内存中。Android 提供了 4 种数据持久化方式,包括 SharedPreferences(偏好设置)、文件存储、SQLite(数据库存储)和 ContentProvider。

10.1 SharedPreferences

Android 提供了一个 SharedPreferences 类,它是一个轻量级的存储类,适合保存用户偏好设置参数和读取设置参数。SharedPreferences 类采用 XML 格式文件存放数据,可以将一些简单的数据类型的数据,包括 boolean 类型、int 类型、float 类型、long 类型以及 String 类型的数据,以键值对的形式存储在应用程序的私有 Preferences 目录(/data/data/<应用包名>/shared_prefs/)中,这种 Preferences 机制广泛应用于存储应用程序中的配置信息。

10.1.1 获取 SharedPreferences 对象

要想使用 SharedPreference 首先要获取 SharedPreference 对象。Android 系统提供了三种获取 SharedPreference 对象的方法。

第一种是通过调用上下文的方法 getSharedPreferences 获取 SharedPreferences 的实例化对象。getSharedPreferences 方法有两个参数,第一个是文件名,即这个偏好设置文件存储的文件名称,第二个是访问文件的模式,有几个固定的常量,分别代表这个文件的访问权限,如 MODE_PRIVATE 表示私有文件,只能被这个应用使用。

第二种是通过调用上下文的方法 getPreferences 获取 SharedPreferences 的实例化对象。getPreferences 方法只有一个参数,表示访问文件的模式,和 getSharedPreferences 方法的第二个参数相同。SharedPreferences 文件名是以当前的 Activity 命名的,所以一个 Activity 只有一个对应的偏好设置文件。

第三种是通过 PreferenceManager 类的静态方法 getDefaultSharedPreferences 来获取 SharedPreference 对象,这个静态方法的参数是应用程序上下文对象。SharedPreferences 文件名是以当前的应用的包名命名的,所以一个应用只对应一个这样的文件。

10.1.2 保存 SharedPreferences

要想保存 SharedPreferences，首先要通过 SharedPreferences 对象的 edit()方法来获取 Editor 对象；然后调用 Editor 对象的 put()方法把数据存放到 SharedPreferences 对象中。put()方法有两个参数，第一个是存储数据的键信息，第二个是存储数据的值信息。最后要调用 Editor 对象的 commit()方法来把数据保存到 XML 文件中，如果不调用 commit()方法数据是不会写入到文件中的。

10.1.3 读取 SharedPreferences

读取 SharedPreferences，要调用 SharedPreferences 对象的 get()方法。get()方法根据读取数据类型的不同，分别有 getString()、getBoolean()等形式。该方法有两个参数，第一个是要读取数据的键信息，第二个是默认返回值，即如果没有读取到相关数据，可以使用这个值作为默认值返回。

10.1.4 SharedPreferences 案例

案例在启动时使用 SharedPreference 写入信息，然后再读取信息，并使用 Log 显示出读取的信息。案例程序包括一个布局文件（activity_main.xml）和一个 Activity 组件类（MainActivity 类）。

案例 Activity 组件 MainActivity 类代码如下。

```java
package com.example.android_demo10_1;

import android.app.Activity;
import android.content.SharedPreferences;
import android.os.Bundle;
import android.preference.PreferenceManager;
import android.util.Log;
/**
 *演示偏好设置类
 */
public class MainActivity extends Activity {

    //声明偏好设置类
    private SharedPreferences mSharedPreferencesFile;
    private SharedPreferences mSharedPreferencesActivity;
    private SharedPreferences mSharedPreferencesApp;

    @Override
    protected void onCreate(Bundle savedInstanceState) {
        super.onCreate(savedInstanceState);
```

```
//设置布局
setContentView(R.layout.activity_main);
//获得偏好设置类对象
mSharedPreferencesFile = getSharedPreferences ( " preference ", MODE_
    PRIVATE);
mSharedPreferencesActivity=getPreferences(MODE_PRIVATE);
mSharedPreferencesApp=
    PreferenceManager.getDefaultSharedPreferences(this);
//获得偏好设置类中的 Editor 对象
SharedPreferences.Editor _EditorFile=mSharedPreferencesFile.edit();
//存放数据
_EditorFile.putString("key01","value01");
//提交数据
_EditorFile.commit();

SharedPreferences.Editor _EditorActivity=
    mSharedPreferencesActivity.edit();
_EditorActivity.putString("key02","value02");
_EditorActivity.commit();

SharedPreferences.Editor _EditorApp=mSharedPreferencesApp.edit();
_EditorApp.putString("key03","value03");
_EditorApp.commit();

//打印日志
Log.i("TAG", mSharedPreferencesFile.getString("key01", ""));
Log.i("TAG",mSharedPreferencesActivity.getString("key02",""));
Log.i("TAG",mSharedPreferencesApp.getString("key03",""));
    }
}
```

案例运行后,LogCat 显示如图 10-1 所示。

Tag	Text
TAG	value01
TAG	value02
TAG	value03

图 10-1 SharedPreferences 案例运行 LogCat 显示

10.2 文件存储

偏好设置虽然使用比较灵活方便,但它有自己的局限性,即只能存放一些简单的键值对信息,对于其他较为复杂的数据就不行了,例如一篇文章。这时就需要使用 Android

的文件存储功能。Android 的文件存储根据存放位置的不同分为内部存储和外部存储两种存储方式。

10.2.1 内部存储

内部存储把要存放的数据文件存储在移动设备内部的存储器中，即/data/data/＜应用包名＞目录中，其实偏好设置就是一种特殊的内部存储。

注意内部存储不是内存。内部存储位于系统中一个很特殊的位置，如果想将一个文件存储于内部存储中，那么这个文件默认只能被你的应用访问，且一个应用所创建的所有文件都在和应用包名相同的目录下。也就是说应用创建于内部存储的文件，与这个应用是关联起来的。当一个应用卸载之后，内部存储中的这些文件也被删除。

从技术上来讲如果在创建内部存储文件的时候将文件属性设置成可读，其他的应用如果知道这个应用的包名，便能够访问这个应用的数据；如果应用创建的文件属性是私有的(private)，那么即使知道包名其他应用也无法访问。

内部存储一般使用 Context 提供的方法来获取输入/输出流并进行操作。相关操作方法如下。

（1）Context.openFileOutput：通过获取输出流，保存文件内容。参数分别为文件名和存储模式。

（2）Context.openFileInput：通过获取输入流，读取文件内容。参数为文件名。

（3）Context.deleteFile：删除指定的文件，参数为将要删除的文件的名称。

（4）Context.fileList：获取文件名列表，为所有文件名数组。

（5）Context.getFilesDir()：获取文件的绝对路径(/data/data/＜应用包名＞/files/filename)，即本身应用的内部存储空间，相当于应用在内部存储上的根目录。

存储模式参数为一个常量，定义如下。

（1）Context.MODE_PRIVATE：为默认操作模式，代表该文件是私有数据，只能被应用本身访问，在该模式下写入的内容会覆盖原文件的内容。

（2）Context.MODE_APPEND：检查文件是否存在，存在就向文件追加内容，否则就创建新文件。

（3）MODE_WORLD_READABLE：表示当前文件可以被其他应用读取。

（4）MODE_WORLD_WRITEABLE：表示当前文件可以被其他应用写入。

在使用模式存储时，可以用＋来选择多种模式，例如"openFileOutput(FILENAME, Context.MODE_PRIVATE＋MODE_WORLD_READABLE);"。

10.2.2 外部存储

内部存储虽然灵活方便，但应用一旦被卸载，存储的文件也就丢失了。如果使用 SD 卡的外部存储器就可以永久地保存相关文件。所有的 Android 设备都有外部存储和内部存储，这两个名称来源于 Android 的早期设备，那个时候的设备内部存储确实是固定的，而外部存储确实是可以像 U 盘一样移动的。但是在后来的设备中，很多移动设备都将自己机身的存储进行了扩展,将存储在概念上分成了"内部"(Internal)和"外部"(External)两部分,但其实都在设备内部。所以不管 Android 设备是否有可移动的 SD 卡,它们总是有外部存储和内部存储。

可以使用 Environment 的静态方法 getExternalStorageDirectory()来获取外部存储的根路径,外部存储中的文件是可以被用户或者其他应用程序修改的。

不是所有的 Android 设备都有外部存储,所以在使用外部存储之前,必须要先检查外部存储的当前状态,以判断是否可用。Environment 的静态方法 getExternalStorageState()可以获取当前外部存储器的状态。

要使用 SD 卡需要添加如下权限:

```xml
<uses-permission android:name="android.permission.MOUNT_UNMOUNT_FILESYSTEMS"/>
<uses-permission android:name="android.permission.READ_EXTERNAL_STORAGE"/>
<uses-permission android:name="android.permission.WRITE_EXTERNAL_STORAGE"/>
```

10.2.3 文件存储案例

案例程序在启动后,先把数据写入到内部存储中,然后从内部存储中读出数据;之后再检测外部存储状态,确定是否可用;如果可用,把数据写入到外部存储中,接着从外部存储中读出刚写入的数据。两次从内、外存储的文件中读取的数据会用 Log 输出显示。

案例中包括一个布局文件(activity_main.xml)和一个 Activity 组件类(MainActivity 类)。Activity 组件 MainActivity 类的代码如下。

```java
package com.example.android_demo10_2;

import java.io.File;
import java.io.FileInputStream;
import java.io.FileOutputStream;
import org.apache.http.util.EncodingUtils;
import android.app.Activity;
import android.os.Bundle;
import android.os.Environment;
import android.os.StatFs;
import android.util.Log;

@SuppressWarnings("deprecation")
public class MainActivity extends Activity {
    //编码格式
    public static final String ENCODING_STR="UTF-8";
    //定义文件名称
    String fileName1="test1.txt";           //写入内部存储文件名
    String fileName2="test2.txt";           //写入外部存储文件名
    //写入和读出的数据信息
    String message="欢迎大家学习 Android!";
```

```java
//是否有 SD 卡
boolean mExternalStorageAvailable=false;
//SD 卡是否可写
boolean mExternalStorageWriteable=false;

@Override
protected void onCreate(Bundle savedInstanceState) {
    super.onCreate(savedInstanceState);
    setContentView(R.layout.activity_main);
    //写入数据到内部存储文件
    writeFileData(fileName1, message);
    //从内部存储文件读出数据
    String result=readFileData(fileName1);
    Log.e("Total Internal Memory Size",""+getTotalInternalMemorySize());
    Log.e("Available Internal Memory Size",""+getAvailableInternalMemorySize());
    Log.e("Write Internal-file Path",getFilesDir().getPath());
    Log.e("Read Internal-file",result);
    //判断外部存储是否可用
    if (externalMemoryAvailable()){
        Log.e("Total External Memory Size",""+getTotalExternalMemorySize());
        Log.e(" Available External Memory Size ","" + getAvailableExternalMemorySize());
        //写入数据到外部存储文件
        writeExtFileData(fileName2, message);
        //从外部存储文件读出数据
        result=readExtFileData(fileName2);
        Log.e(" Write External - file Path ", Environment.getExternalStorageDirectory().getAbsolutePath());
        Log.e("Read External-file",result);
    }
}

//向指定的内部存储文件中写入指定的数据
public void writeFileData(String filename, String message) {
    try {
        //获得 FileOutputStream
        FileOutputStream fout=openFileOutput(filename, MODE_PRIVATE);
        //将要写入的字符串转换为 byte 数组
        byte[] bytes=message.getBytes();
        fout.write(bytes);              //将 byte 数组写入文件
        fout.close();                   //关闭文件输出流
    } catch (Exception e) {
        e.printStackTrace();
    }
}
```

```java
        }
        //打开内部存储的指定文件,读取其数据,返回字符串对象
        public String readFileData(String fileName) {
            String result="";
            try {
                FileInputStream fin=openFileInput(fileName);
                //获取文件长度
                int lenght=fin.available();
                byte[] buffer=new byte[lenght];
                fin.read(buffer);
                //将byte数组转换成指定格式的字符串
                result=EncodingUtils.getString(buffer, ENCODING_STR);
                fin.close();
            } catch (Exception e) {
                e.printStackTrace();
            }
            return result;
        }

        public long getAvailableInternalMemorySize(){
            //文件路径
            File path=Environment.getDataDirectory();
            StatFs stat=new StatFs(path.getPath());
            long blockSize=stat.getBlockSize();
            long availableBlocks=stat.getAvailableBlocks();
            return availableBlocks * blockSize;
        }

        public long getTotalInternalMemorySize(){
            File path=Environment.getDataDirectory();
            StatFs stat=new StatFs(path.getPath());
            long blockSize=stat.getBlockSize();
            long totalBlocks=stat.getBlockCount();
            return totalBlocks * blockSize;
        }

        public boolean externalMemoryAvailable(){
            return Environment. getExternalStorageState ( ). equals (Environment.
            MEDIA_MOUNTED);
        }

        public long getAvailableExternalMemorySize(){
            if(externalMemoryAvailable()){
```

```java
            File path=Environment.getExternalStorageDirectory();
            StatFs stat=new StatFs(path.getPath());
            long blockSize=stat.getBlockSize();
            long availableBlocks=stat.getAvailableBlocks();
            return availableBlocks * blockSize;
        }
        else{
            return -1;
        }
    }

    public long getTotalExternalMemorySize(){
        if(externalMemoryAvailable()){
            File path=Environment.getExternalStorageDirectory();
            StatFs stat=new StatFs(path.getPath());
            long blockSize=stat.getBlockSize();
            long totalBlocks=stat.getBlockCount();
            return totalBlocks * blockSize;
        }
        else{
            return -1;
        }
    }

    //向指定的外部存储文件中写入指定的数据
    public void writeExtFileData(String filename, String message) {
        try {
            File file = new File (Environment.getExternalStorageDirectory(),
            "/"+filename);
            FileOutputStream fos=new FileOutputStream(file);
            //将要写入的字符串转换为 byte 数组
            byte[] bytes=message.getBytes();
            fos.write(bytes);
            fos.flush();
            fos.close();
        } catch (Exception e) {
            e.printStackTrace();
        }
    }

    //打开外部存储的指定文件,读取其数据,返回字符串对象
    public String readExtFileData(String fileName) {
        String result="";
        try {
```

```
            File file= new File (Environment.getExternalStorageDirectory(),
            "/"+fileName);
            FileInputStream fin=new FileInputStream(file);
            //获取文件长度
            int lenght=fin.available();
            byte[] buffer=new byte[lenght];
            fin.read(buffer);
            //将 byte 数组转换成指定格式的字符串
            result=EncodingUtils.getString(buffer, ENCODING_STR);
            fin.close();
        } catch (Exception e) {
            e.printStackTrace();
        }
        return result;
    }
}
```

案例运行 LogCat 显示如图 10-2 所示。

Tag	Text
Total Internal Memory Size	5049040896
Available Internal Memory Size	579014656
Write Internal-file Path	/data/data/com.example.android_demo10_2/files
Read Internal-file	欢迎大家学习Android!
Total External Memory Size	5049040896
Available External Memory Size	579014656
Write External-file Path	/storage/emulated/0
Read External-file	欢迎大家学习Android!

图 10-2 文件存储案例运行 LogCat 显示

10.3 SQLite 数据库存储

10.3.1 SQLite 简介

Android 系统中内置了一个 SQLite 数据库，Android 运行时环境包含了完整的 SQLite。SQLite 是轻量级嵌入式数据库引擎，它支持 SQL 语言，基本上符合 SQL-92 标准，并且只利用很少的内存就有很好的性能。SQLite 是开源的，任何人都可以使用它，许多开源项目都使用了 SQLite。SQLite 由 SQL 编译器、内核、后端以及附件组成，利用虚拟机和虚拟数据库引擎（VDBE），使调试、修改和扩展 SQLite 的内核变得更加方便。

SQLite 和其他数据库最大的不同就是对数据类型的支持，例如创建一个表时，可以在 CREATE TABLE 语句中指定某列的数据类型，但是可以把任何数据类型放入任何列中；当把某个值插入数据库时，SQLite 将检查它的类型，如果该类型与关联的列不匹配，SQLite 会尝试将该值转换成该列的类型；如果不能转换，则该值将作为其本身具有的类型存储，SQLite 称这种类型为"弱类型"。此外，SQLite 不支持一些标准的 SQL 功能，特别是外键约束（FOREIGN KEY），嵌套 Transcaction 和 RIGHT OUTER JOIN 及 FULL OUTER JOIN，还有一些 ALTER TABLE 功能。除了上述功能外，SQLite 是一个完整的 SQL 系统，拥有完整的触发器等。

SQLite 数据库默认存储在/data/data/＜应用包名＞/databases/目录下（特殊情况，应用程序也可以将其放在外部存储）。

10.3.2 SQLiteOpener

Android 提供了 SQLiteOpenHelper 帮助用户创建一个数据库，只要继承 SQLiteOpenHelper 类，就可以轻松地创建数据库。SQLiteOpenHelper 类根据开发应用程序的需要，封装了创建和更新数据库使用的逻辑。

SQLiteOpenHelper 的子类，至少需要实现三个方法。

（1）构造函数：调用父类 SQLiteOpenHelper 的构造函数。这个方法需要 4 个参数即上下文环境（Content）、数据库名字、一个可选的 Cursor（通常是 Null）、一个代表正在使用的数据库模型版本的整数。

（2）onCreate()方法：需要一个 SQLiteDatabase 对象作为参数，根据需要对这个对象填充表和初始化数据。

（3）onUpgrage()方法：需要三个参数，即一个 SQLiteDatabase 对象、一个旧的版本号和一个新的版本号，这样就可以清楚如何把一个数据库从旧的模型转变到新的模型。

在创建构造方法时就会创建数据库，一般可以在 Activity 的 onCreate()方法中执行 SQL 语句来创建表。如果数据库的版本号发生变化，就会调用 onUpgrage()方法，来执行 SQL 语句对表进行修改。

10.3.3 数据库操作

使用 SQLiteDatabase 对象可以进行数据库操作，其方法有如下几个。

（1）insert(String table, String nullColumnHack, ContentValues values)；

向数据库表中写入一条数据。

（2）update（String table, Contentvalues values, String whereClause, String [] whereArgs)；

更新一条数据。

（3）delete(String table, String whereClause, String [] whereArgs)；

删除一条数据。

（4）Cursor query(String table, String[] columns, String whereClause, String [] whereArgs, String groupBy, String having, String orderBy, String limit)；

查询数据，返回值为 Cursor，将查询到的结果都存在 Cursor。

上面的方法中，其参数表示介绍如下。

(1) table：数据库表的名称。
(2) columns：数据库列名称数组。
(3) whereClause：查询条件，表示 WHERE 表达式，比如"age>? and age<?"。
(4) whereArgs：占位符的实际参数值。
(5) groupBy：指定分组的列名。
(6) having：指定分组条件。
(7) orderBy：指定排序的列名。
(8) limit：分页查询限制。
(9) nullColumnHack：表示如果插入的数据每一列都为空的话，需要指定此行中某一列的名称，系统将此列设置为 NULL，不至于出现错误。
(10) ContentValues：键值对组成的 Map，key 代表列名，value 代表该列要插入的值。

query 方法的返回值 Cursor 是一个游标接口，每次查询的结果都会保存在 Cursor 中，可以通过遍历 Cursor 的方法得到当前查询到的所有信息。Cursor 的方法如下。

(1) moveToFirst()：将 Curor 的游标移动到第一条。
(2) moveToLast()：将 Curor 的游标移动到最后一条。
(3) move(int offset)：将 Curor 的游标移动到指定 ID。
(4) moveToNext()：将 Curor 的游标移动到下一条。
(5) moveToPrevious()：将 Curor 的游标移动到上一条。
(6) getCount()：得到 Cursor 总记录条数。
(7) isFirst()：判断当前游标是否为第一条记录。
(8) isLast()：判断当前游标是否为最后一条数据。
(9) getInt(int columnIndex)：根据列名称获得列索引 ID。
(10) getString(int columnIndex)：根据索引 ID 得到表中的字段。

10.3.4 SQLite 案例

本案例实现了对 SQLite 数据库的基本操作，包括创建/打开数据库，对表进行添加、更新、删除和查询操作。

案例程序包括一个菜单布局文件(menu.xml)，一个存储数据库名称、表名及字段名的类 DBinfo，还包括 Activity 组件类(MainActivity 类)及其布局文件。

1. 布局文件 activity_main.xml

```
<?xml version="1.0" encoding="utf-8"?>
<LinearLayout
    xmlns:android="http://schemas.android.com/apk/res/android"
    android:orientation="vertical"
    android:layout_width="fill_parent"
    android:layout_height="fill_parent">
```

```xml
<LinearLayout
    android:layout_width="fill_parent"
    android:layout_height="wrap_content"
    android:orientation="horizontal">

    <TextView
        android:layout_width="wrap_content"
        android:layout_height="wrap_content"
        android:layout_margin="5dp"
        android:text="ID:"
        android:textSize="20sp" />

    <EditText
        android:id="@+id/edit1"
        android:layout_width="200dp"
        android:layout_height="wrap_content"
        android:layout_margin="5dp"
        android:textSize="20sp" />

</LinearLayout>

<LinearLayout
    android:layout_width="fill_parent"
    android:layout_height="wrap_content"
    android:gravity="center"
    android:orientation="vertical">

    <TextView
        android:id="@+id/view"
        android:layout_width="fill_parent"
        android:layout_height="wrap_content"
        android:layout_margin="5dp"
        android:hint="记录内容"
        android:textColor="#FFFFCC"
        android:textSize="20sp" />

    <EditText
        android:id="@+id/edit2"
        android:layout_width="fill_parent"
        android:layout_height="100dp"
        android:layout_margin="5dp"
        android:background="#aaffff"
```

```xml
            android:scrollbars="vertical"
            android:textSize="20sp" />

    </LinearLayout>

    <ListView
        android:id="@+id/lv"
        android:layout_width="fill_parent"
        android:layout_height="match_parent"
        android:layout_gravity="start"
        android:layout_margin="3dp"
        android:background="#eeffee" />

</LinearLayout>
```

2. 菜单布局文件 menu.xml

```xml
<?xml version="1.0" encoding="utf-8"?>
<menu xmlns:android="http://schemas.android.com/apk/res/android">
<item android:id="@+id/save" android:title="保存"/>
<item android:id="@+id/query" android:title="查询"/>
<item android:id="@+id/update" android:title="修改"/>
<item android:id="@+id/delete" android:title="删除"/>
</menu>
```

3. DBinfo.java

```java
package com.example.android_demo10_3;

public class DBinfo {
    public static final String _DBNAME="notes";
    public static final String _TABLENAME="mark";
    public static final String _NO="id";
    public static final String _CONTENT="txt";
    public static final String _CREATTIME="ctime";
    public static final String _LASTTIME="ltime";
}
```

4. Activity 组件 MainActivity 类 MainActivity.java

```java
package com.example.android_demo10_3;

import jva.text.SimpleDateFormat;
```

```java
import java.util.ArrayList;
import java.util.Date;
import java.util.List;
import android.annotation.SuppressLint;
import android.app.Activity;
import android.content.ContentValues;
import android.content.Context;
import android.database.Cursor;
import android.database.sqlite.SQLiteDatabase;
import android.os.Bundle;
import android.util.Log;
import android.view.Menu;
import android.view.MenuInflater;
import android.view.MenuItem;
import android.widget.ArrayAdapter;
import android.widget.EditText;
import android.widget.ListView;
import android.widget.Toast;

public class MainActivity extends Activity {
    private SQLiteDatabase db;
    private ListView lv;
    @SuppressLint("SimpleDateFormat")
    private static final SimpleDateFormat sdf=new SimpleDateFormat(
            "yyyy-MM-dd hh:mm:ss");

    @Override
    public void onCreate(Bundle savedInstanceState) {
        super.onCreate(savedInstanceState);
        setContentView(R.layout.activity_main);
        db=openOrCreateDatabase(DBinfo._DBNAME, Context.MODE_PRIVATE, null);
        int version=db.getVersion();
        //创建数据库
        if (version<1) {
            db.execSQL("create table "+DBinfo._TABLENAME+"("+DBinfo._NO
                    +" text primary key,"+DBinfo._CONTENT+" text,"
                    +DBinfo._CREATTIME+" text,"+DBinfo._LASTTIME+" text)");
            db.setVersion(1);
        }
        //通过id获取界面控件ListView
        lv=(ListView) findViewById(R.id.lv);
        initview();
    }
```

```java
private void initview(){
    List<String>data=new ArrayList<String>();
    try{
        //查询表中所有数据
        Cursor cursor=db.query(DBinfo._TABLENAME,
            null,null,null,null,null);
        if(cursor.getCount()>0){
            while(cursor.moveToNext()){
                String record="ID: "
                    +cursor.getString(cursor.getColumnIndex(DBinfo._NO))
                    +"\n记录内容: "+cursor
                    .getString(cursor.getColumnIndex(DBinfo._CONTENT));
                data.add(record);
            }
        }else {
            data.add("没有记录");
        }
    }catch(Exception e) {
        Log.e("db-error",e.getMessage());
    }
    //给 ListView 设置适配器,显示表中所有数据
    lv.setAdapter(new ArrayAdapter<String>(this,
        android.R.layout.simple_list_item_1,data));
}
@Override
public boolean onCreateOptionsMenu(Menu menu) {
    MenuInflater flater=getMenuInflater();
    flater.inflate(R.menu.menu, menu);
    return super.onCreateOptionsMenu(menu);
}

@Override
public boolean onOptionsItemSelected(MenuItem item) {

    EditText edit1=(EditText) findViewById(R.id.edit1);
    EditText edit2=(EditText) findViewById(R.id.edit2);

    String id=edit1.getText().toString();
    String content=edit2.getText().toString();
    int itemId=item.getItemId();

    switch (itemId) {
```

```java
            //保存信息
        case R.id.save:
            db.execSQL("insert into "+DBinfo._TABLENAME+"("
                +DBinfo._NO+","+DBinfo._CONTENT+","
                +DBinfo._CREATTIME+","
                +DBinfo._LASTTIME+") values(?,?,?,?)", new Object[]{
                id, content, sdf.format(new Date()),
                sdf.format(new Date())});
            Toast.makeText(this, "保存成功", Toast.LENGTH_LONG).show();
            break;

            //查询信息
        case R.id.query:
            Cursor cursor=db.query(DBinfo._TABLENAME,
                new String[]{DBinfo._CONTENT},DBinfo._NO+"=?",
                new String[]{id},null,null,null);
            int index=cursor.getColumnIndex(DBinfo._CONTENT);
            if(cursor.moveToNext()){
                edit2.setText(cursor.getString(index));
            }else{
                Toast.makeText(this, "无此记录", Toast.LENGTH_LONG).show();
            }
            break;

            //修改信息
        case R.id.update:
            ContentValues values=new ContentValues();
            values.put(DBinfo._CONTENT,content);
            values.put(DBinfo._LASTTIME,sdf.format(new Date()));
            db.update(DBinfo._TABLENAME,values,DBinfo._NO+"=?",
                new String[]{id});
            Toast.makeText(this,"修改成功",Toast.LENGTH_LONG).show();
            break;

            //删除信息
        case R.id.delete:
            db.delete(DBinfo._TABLENAME,DBinfo._NO+"=?",new String[]{id});
            Toast.makeText(this,"删除成功",Toast.LENGTH_LONG).show();
            break;
        }
        initview();
        return super.onOptionsItemSelected(item);
    }
```

```
@Override
protected void onDestroy() {
    //TODO Auto-generated method stub
    db.close();
    db=null;
    super.onDestroy();
}
```

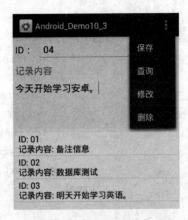

图 10-3 SQLite 案例运行效果

案例运行效果如图 10-3 所示。案例程序开始会在 Activity 底部的 ListView 中显示数据库表中的所有记录,用户单击菜单可以实现对表的添加(保存)、查询、更新(修改)及删除操作。

10.4 ContentProvider

10.4.1 ContentProvider 简介

ContentProvider(内容提供者)为存储和获取数据提供了统一的接口,使用 ContentProvider 可以在不同的应用程序之间共享数据。Android 使用 ContentProvider 为一些数据提供外部访问的接口,例如通话记录、联系人、短信、多媒体文件等,以方便开发人员访问数据,减少开发成本。

ContentProvider 将其数据用数据库中简单的表模型展示,无论数据的来源是什么,一行表示一条记录。ID 可用于匹配相关联的表。例如,在一个表中找到联系人的电话号码,在另一张表中找到这个联系人的头像。

所有的 ContentProvider 实现一个通用的接口用于查询(query)并返回结果,当然也有增加(add)、修改(update)和删除(delete)数据的通用接口。

一般通过 ContentResolver 间接地获取 ContentProvider 内容、在 Activity 中可以通过 getContentResolver()方法获取 ContentResolver。ContentResolver 提供的接口和 ContentProvider 中需要实现的接口对应。Android 系统负责初始化所有的 ContentProvider,不需要用户自己创建。实际上,ContentProvider 的用户都不可能直接访问到 ContentProvider 实例,只能通过 ContentResolver 在中间代理。

查询时,Android 系统识别其查询目标 ContentProvider,确保其启动和运行。一般,每一个 ContentProvider 都只有一个实例(单例模式),但可以与多个在不同应用或者进程中的 ContentResolver 进程通信。查询返回的结果为 Cursor(游标),可以在记录或者字段之间游动。

10.4.2 访问手机数据信息

Android 所提供的 ContentProvider 都存放在 android.provider 包当中。每一个

ContentProvider 都拥有一个公共的 URI,这个 URI 用于表示这个 ContentProvider 所提供的数据。URI 主要包含了两部分信息:需要操作的 ContentProvider,对 ContentProvider 中的什么数据进行操作。

一个 URI 由以下几部分组成。

(1) scheme:ContentProvider 的 scheme 已经由 Android 规定为 content://。

(2) 主机名(或 Authority):用于唯一标识这个 ContentProvider,外部调用者可以根据这个标识来找到它。

(3) 路径(path):可以用来表示要操作的数据,路径的构建应根据业务而定,举例如下。

① 要操作 contact 表中 id 为 10 的记录,可以构建这样的路径/contact/10;

② 要操作 contact 表中 id 为 10 的记录的 name 字段,contact/10/name;

③ 要操作 contact 表中的所有记录,可以构建这样的路径/contact。

要操作的数据不一定来自数据库,也可以是文件等其他存储方式。如果要把一个字符串转换成 URI,可以使用 URI 类中的 parse()方法,例如:

```
Uri uri = Uri.parse("content://com.example.provider.contactprovider/contact");
```

实现 ContentProvider 的过程如下。

(1) 定义一个 CONTENT_URI 常量;

(2) 定义一个类,继承 ContentProvider;

(3) 实现 query、insert、update、delete、getType、onCreate 方法;

(4) 在 AndroidManifest.xml 当中进行声明。

通常程序中会使用系统的 URI 来访问系统信息,通过使用 ContentResolver.query()方法来查询系统数据信息,其格式如下:

```
Cursor android.content.ContentResolver.query(Uri uri, String[] projection, String selection, String[] selectionArgs, String sortOrder)
```

query()方法会根据给定的 URI 查询,返回 Cursor 结果集,其参数含义如下。

(1) uri:要查询数据的 URI;

(2) projection:要查询的信息,null 表示查询所有信息;

(3) selection:过滤条件,如同 SQL 语句一样使用"?"做参数;

(4) selectionArgs:以数组的形式表示,"?"做参数;

(5) sortOrder:返回结果集的排序方式。

10.4.3 ContentProvider 案例

案例中界面有三个按钮,单击后分别启动不同的 Activity,每个 Activity 分别在其界面的 ListView 及 Log 中显示系统联系人、系统电话记录和系统短信记录。

案例程序包括 4 个 Activity 组件类及其布局文件,还有一个 ListView 的条目(Item)

布局文件。

1. 布局文件 activity_main.xml

```xml
<LinearLayout xmlns:android="http://schemas.android.com/apk/res/android"
    xmlns:tools="http://schemas.android.com/tools"
    android:id="@+id/LinearLayout1"
    android:layout_width="match_parent"
    android:layout_height="match_parent"
    android:orientation="vertical">

    <Button
        android:id="@+id/contact_btn"
        android:layout_width="fill_parent"
        android:layout_height="wrap_content"
        android:text="Contact" />

    <Button
        android:id="@+id/calllog_btn"
        android:layout_width="fill_parent"
        android:layout_height="wrap_content"
        android:text="CallLog" />

    <Button
        android:id="@+id/smslog_btn"
        android:layout_width="match_parent"
        android:layout_height="wrap_content"
        android:text="SMSLog" />

</LinearLayout>
```

2. 布局文件 activity_calllog.xml

```xml
<?xml version="1.0" encoding="utf-8"?>
<LinearLayout xmlns:android="http://schemas.android.com/apk/res/android"
    android:layout_width="match_parent"
    android:layout_height="match_parent"
    android:orientation="vertical">

    <ListView
        android:id="@+id/calllog_lv"
        android:layout_width="fill_parent"
        android:layout_weight="1.0"
        android:layout_height="0dp"/>
```

</LinearLayout>

3. 布局文件 activity_contacts.xm

```xml
<?xml version="1.0" encoding="utf-8"?>
<LinearLayout xmlns:android="http://schemas.android.com/apk/res/android"
    android:layout_width="match_parent"
    android:layout_height="match_parent"
    android:orientation="vertical">

    <ListView
        android:id="@+id/contacts_lv"
        android:layout_width="fill_parent"
        android:layout_weight="1.0"
        android:layout_height="0dp"/>

</LinearLayout>
```

4. 布局文件 activity_smslog.xml

```xml
<?xml version="1.0" encoding="utf-8"?>
<LinearLayout xmlns:android="http://schemas.android.com/apk/res/android"
    android:layout_width="match_parent"
    android:layout_height="match_parent"
    android:orientation="vertical">

    <ListView
        android:id="@+id/smslog_lv"
        android:layout_width="fill_parent"
        android:layout_weight="1.0"
        android:layout_height="0dp"/>

</LinearLayout>
```

5. 布局文件 item.xml

```xml
<?xml version="1.0" encoding="utf-8"?>
<LinearLayout xmlns:android="http://schemas.android.com/apk/res/android"
    android:layout_width="match_parent"
    android:layout_height="match_parent"
    android:orientation="horizontal">

    <ImageView
        android:id="@+id/img"
```

```xml
        android:layout_width="wrap_content"
        android:layout_height="wrap_content"
        android:layout_gravity="center_vertical"
        android:layout_margin="5dp"
        android:src="@drawable/itemicon" />

    <TextView
        android:id="@+id/txt"
        android:layout_width="wrap_content"
        android:layout_height="wrap_content"
        android:text="" />

</LinearLayout>
```

6. Activity 组件 MainActivity 类

```java
package com.tarena.demo_10_4;

import android.app.Activity;
import android.content.Intent;
import android.os.Bundle;
import android.view.View;
import android.view.View.OnClickListener;

public class MainActivity extends Activity {

    @Override
    protected void onCreate(Bundle savedInstanceState) {
        super.onCreate(savedInstanceState);
        //设置布局
        setContentView(R.layout.activity_main);
        //联系人按钮
        findViewById(R.id.contact_btn).setOnClickListener(new OnClickListener() {

            @Override
            public void onClick(View arg0) {
                //跳转到联系人界面
                Intent intent=new Intent();
                intent.setClass(MainActivity.this, ContactsActivity.class);
                startActivity(intent);
            }
        });
        //通话记录按钮
        findViewById(R.id.calllog_btn).setOnClickListener(new OnClickListener() {
```

```java
            @Override
            public void onClick(View arg0) {
                //跳转到通话记录界面
                Intent intent=new Intent();
                intent.setClass(MainActivity.this, CallLogActivity.class);
                startActivity(intent);
            }
        });
        //短信按钮
        findViewById(R.id.smslog_btn).setOnClickListener(new OnClickListener() {

            @Override
            public void onClick(View arg0) {
                //跳转到短信界面
                Intent intent=new Intent();
                intent.setClass(MainActivity.this, SMSLogActivity.class);
                startActivity(intent);
            }
        });
    }

}
```

7. Activity 组件 CallLogActivity 类

```java
package com.tarena.demo_10_4;

import java.util.ArrayList;
import java.util.HashMap;
import java.util.List;
import java.util.Map;

import android.app.Activity;
import android.database.Cursor;
import android.os.Bundle;
import android.provider.CallLog;
import android.util.Log;
import android.widget.ListView;
import android.widget.SimpleAdapter;

public class CallLogActivity extends Activity {

    @Override
```

```java
protected void onCreate(Bundle savedInstanceState) {
    super.onCreate(savedInstanceState);
    setContentView(R.layout.activity_calllog);
    //通过 id 获取界面控件 ListView
    ListView lv=(ListView) findViewById(R.id.calllog_lv);
    //给 ListView 设置适配器 SimpleAdapter
    lv.setAdapter(new SimpleAdapter(this, queryCallLog(), R.layout.item,
        new String[] { "nametext", "namepic" },
        new int[] { R.id.txt, R.id.img}));
}
/**
 * 查询通话记录
 */
private List<Map<String, Object>>queryCallLog(){
    //创建数据源集合
    List<Map<String, Object>>data=new ArrayList<Map<String, Object>>();
    //获得通话记录 Cursor
    Cursor cursor=getContentResolver().query(CallLog.Calls.CONTENT_URI,
        null, null, null, CallLog.Calls.DEFAULT_SORT_ORDER);
    //判断 Cursor 是否为 NULL 或者是否有数据
    if(cursor !=null && cursor.getCount()>0){
        //循环读取数据
        while (cursor.moveToNext()) {
            String phoneName=cursor.getString(cursor
                .getColumnIndex(CallLog.Calls.NUMBER));
            String name=cursor.getString(cursor
                .getColumnIndex(CallLog.Calls.CACHED_NAME));
            String type=cursor.getString(cursor
                .getColumnIndex(CallLog.Calls.TYPE));
            String date=cursor.getString(cursor
                .getColumnIndex(CallLog.Calls.DATE));
            //日志显示
            Log.i("info", "PhoneName-->"+phoneName+",Name-->"+name+",
                Type-->"+type+",Date-->"+date);
            //创建集合中的数据 Map
            Map<String, Object>map1=new HashMap<String, Object>();
            map1.put("nametext", "PhoneName: "+phoneName+",
                Name: "+name+"\nType: "+type+", Date: "+date);
            map1.put("namepic",R.drawable.itemicon);
            //将 Map 数据添加到集合中
            data.add(map1);
        }
        //关闭游标
        cursor.close();
```

```
        }
        return data;
    }
}
```

8. Activity 组件 ContactsActivity 类

```java
package com.tarena.demo_10_4;

import java.util.ArrayList;
import java.util.HashMap;
import java.util.List;
import java.util.Map;

import android.app.Activity;
import android.database.Cursor;
import android.os.Bundle;
import android.provider.ContactsContract.CommonDataKinds.Phone;
import android.util.Log;
import android.widget.ListView;
import android.widget.SimpleAdapter;

public class ContactsActivity extends Activity {

    @Override
    protected void onCreate(Bundle savedInstanceState) {
        //TODO Auto-generated method stub
        super.onCreate(savedInstanceState);
        setContentView(R.layout.activity_contacts);
        //通过 id 获取界面控件 ListView
        ListView lv=(ListView) findViewById(R.id.contacts_lv);
        //给 ListView 设置适配器 SimpleAdapter
        lv.setAdapter(new SimpleAdapter(this, queryContent(), R.layout.item,
                new String[] { "nametext", "namepic" },
                    new int[] { R.id.txt, R.id.img}));
    }
    /**
     * 查询通讯录
     */
    private List<Map<String, Object>> queryContent(){
        //创建数据源集合
        List<Map<String, Object>> data=new ArrayList<Map<String, Object>>();
        //获得联系人游标对象
        Cursor cursor=getContentResolver().query(Phone.CONTENT_URI, null,
```

```java
                null, null, null);
        //判断游标是否为 NULL 或者数据是否为 0
        if(cursor !=null && cursor.getCount()>0 ){
            //循环数据游标
            while (cursor.moveToNext()) {
                //获得数据
                int _id=cursor.getInt(cursor.getColumnIndex(Phone._ID));
                String name=cursor.getString(cursor
                    .getColumnIndex(Phone.DISPLAY_NAME));
                String phoneNumber=cursor.getString(cursor
                    .getColumnIndex(Phone.NUMBER));
                //日志显示
                Log.i("info","_ID-->"+_id+",Name-->"+name+",
                    Number-->"+phoneNumber);
                //创建集合中的数据 Map
                Map<String, Object>map1=new HashMap<String, Object>();
                map1.put("nametext", "ID: "+_id+",
                    Name: "+name+"\nNumber: "+phoneNumber);
                map1.put("namepic",R.drawable.itemicon);
                //将 Map 数据添加到集合中
                data.add(map1);
            }
            //关闭游标
            cursor.close();
        }
        return data;
    }
}
```

9. Activity 组件 SMSLogActivity 类

```java
package com.tarena.demo_10_4;

import java.util.ArrayList;
import java.util.HashMap;
import java.util.List;
import java.util.Map;

import android.app.Activity;
import android.database.Cursor;
import android.net.Uri;
import android.os.Bundle;
import android.util.Log;
import android.widget.ListView;
```

```
    private void addListener() {
        button.setOnClickListener(new OnClickListener() {
            @Override
            public void onClick(View v) {
                InputStream stream=   【1】
                Scanner scanner=new Scanner(stream);
                StringBuffer buffer=new StringBuffer();
                while (    【2】    ) {
                    buffer.append(scanner.next()+"\n");
                }
                scanner.close();
                tvMsg.setText(buffer);
            }
        });
    }
    private void setupView() {
        tvMsg=(TextView) findViewById(R.id.tvMsg);
        button=(Button) findViewById(R.id.button1);
    }
}
```

2. 在 Android 中，如何通过程序把文件存放到 SD 卡中，请写出简要步骤及主要代码。

3. Android 有几种数据存储方式？

4. Android 系统提供了哪几种方法，可以获取 SharedPreference 对象？

5. 使用 SQLiteDatabase 可以进行哪些数据库操作？说明 SQLiteDatabase 的常用方法。

6. ContentProvider 是如何实现数据共享的？

第 11 章

Android 网络编程

在 HTTP 协议的基础上,Android 中提供了两种 HTTP 通信编程方式,分别是直接通信的 HttpURLConnection 接口和附加了用户登录等 Cookie 信息的 HttpClient 接口。

11.1 URL 统一资源定位符

在介绍 URLConnection 接口之前,需要说明 URL 是什么。URL(Uniform Resource Locator)被称为统一资源定位符,也可以被称为网页地址,是因特网上标准的资源地址。通常情况下,URL 由传送协议、服务器、端口号、资源路径组成。Android 中支持的传送协议有 FILE、FTP、HTTP、HTTPS 和 Jar 等。URL 的一般形式是:

```
<URL的传送协议>://<主机>:<端口>/<路径>
```

11.2 使用 URLConnection 接口

对 Java 网络编程熟悉的读者,肯定不会对 URLConnection 感到陌生。URLConnection 属于 Java API 的标准接口,包含在包 java.net 中。而 Android 平台支持 java.net 包中的 API。

通过 URL 中的 openConnection()方法可以获得 URLConnection 对象,该对象表示应用程序与 URL 之间的通信。通过 URLConnection 实例向 URL 发送请求,读取 URL 资源。

通常使用 URLConnection 的步骤如下:
(1) 创建 URL 对象;
(2) 通过调用 URL 对象的 openConnection()方法来创建对象;
(3) 设置 URLConnection 的参数;
(4) 使用 URLConnection 的 getInputStream()获得输入流;
(5) 对输入流进行相应的处理。

11.3 案例 URLConnection

11.3.1 案例功能描述

案例中的界面有一个按钮和一个图片视图,单击按钮后建立网络连接,获取网络图片,更新界面图片。

11.3.2 案例程序结构

案例中包括一个布局文件(activity_main.xml),一个 Activity 组件类(MainActivity 类),用于实现用户界面交互功能。

11.3.3 案例的实现步骤和思路

(1) 创建 Android 项目。
(2) 在 res 目录下的 layout 子目录中创建布局文件 activity_main.xml。
(3) 编写 MainActivity 文件。

在 onCreate 方法中,先调用父类的 onCreate 方法;使用 setContentView 方法加载布局文件 activity_main.xml;获取界面按钮控件,并设置按钮监听器,在监听器中开启新线程,使用 URLConnection 建立网络连接、获取网络图片,完成后通知系统更新界面图片。

11.3.4 案例参考代码

(1) 布局文件 activity_main.xml 代码如下。

```xml
<RelativeLayout xmlns:android="http://schemas.android.com/apk/res/android"
    xmlns:tools="http://schemas.android.com/tools"
    android:layout_width="match_parent"
    android:layout_height="match_parent"
    android:paddingBottom="@dimen/activity_vertical_margin"
    android:paddingLeft="@dimen/activity_horizontal_margin"
    android:paddingRight="@dimen/activity_horizontal_margin"
    android:paddingTop="@dimen/activity_vertical_margin"
    tools:context=".MainActivity">

    <ImageView
        android:id="@+id/iv_image"
        android:layout_width="200dp"
        android:layout_height="200dp"
        android:layout_centerHorizontal="true"
```

```xml
            android:layout_centerVertical="true"
            android:src="@drawable/ic_launcher"/>

    <Button
        android:id="@+id/btn_update_image"
        android:layout_width="wrap_content"
        android:layout_height="wrap_content"
        android:layout_alignParentLeft="true"
        android:layout_alignParentRight="true"
        android:layout_below="@+id/iv_image"
        android:text="更新图片" />

</RelativeLayout>
```

(2) Activity 组件 MainActivity 类代码如下。

```java
package com.example.android_demo11_1;

import java.io.InputStream;
import java.net.URL;
import java.net.URLConnection;
import android.app.Activity;
import android.graphics.Bitmap;
import android.graphics.BitmapFactory;
import android.os.Bundle;
import android.os.Handler;
import android.os.Message;
import android.view.View;
import android.view.View.OnClickListener;
import android.widget.Button;
import android.widget.ImageView;

public class MainActivity extends Activity {

    private ImageView imageIV;
    private Button updateBtn;
    //Handler 通信类
    private Handler handler=new Handler(){
        public void handleMessage(android.os.Message msg) {
            imageIV.setImageBitmap((Bitmap)msg.obj);
        };
    };
```

```java
@Override
protected void onCreate(Bundle savedInstanceState) {
    super.onCreate(savedInstanceState);
    setContentView(R.layout.activity_main);
    //获得控件对象
    imageIV=(ImageView) this.findViewById(R.id.iv_image);
    updateBtn=(Button) this.findViewById(R.id.btn_update_image);
    updateBtn.setOnClickListener(new OnClickListener() {

        @Override
        public void onClick(View arg0) {
            new Thread(){
                public void run() {
                    try {
                        //声明统一资源定位符
                        URL url=
                        new URL("http://192.168.168.41:8080/image/image01.jpg");
                        //获得 URLConnection 对象
                        URLConnection conn=url.openConnection();
                        //设置访问超时
                        conn.setConnectTimeout(3*1000);
                        //获得输入流
                        InputStream ips=conn.getInputStream();
                        //获取图片
                        Bitmap bitmap=BitmapFactory.decodeStream(ips);
                        Message msg=Message.obtain(handler, 0, bitmap);
                        msg.sendToTarget();
                    } catch (Exception e) {
                        //TODO Auto-generated catch block
                        e.printStackTrace();
                    }
                }
            };
        }.start();
        }
    });
}
```

11.3.5 案例运行效果

案例 URLConnection 的运行效果如图 11-1 所示。

图 11-1　案例 URLConnection 运行效果

11.4　使用 HttpClient 接口

　　Apache 开源组织提供了一个 HttpClient 项目，它是简单的 HTTP 客户端，用于发送 HTTP 请求、接收 HTTP 响应。

　　HttpClient 发送请求、接收响应的步骤如下：
　　(1) 创建 HttpClient 对象；
　　(2) 创建 HttpGet 对象或 HttpPost 对象；
　　(3) 使用 HttpGet 对象或 HttpPost 对象的 setEntity()方法，添加请求参数；
　　(4) 使用 HttpClient 对象的 execute()方法发送请求，该方法返回 HttpResponse；
　　(5) 使用 HttpResponse 的 getEntity 方法获得服务器响应。

11.5　案例 HttpClient 接口

11.5.1　案例功能描述

　　案例中界面有一个按钮、两个输入框、一个 ListView，单击按钮后建立网络连接，获取网络 JSON 数据，把数据用 ListView 显示。

11.5.2　案例程序结构

　　案例中包括两个布局文件(activity_main.xml、item.xml)，一个 Activity 组件类

（MainActivity 类），用于实现用户界面交互功能。Java 类 GlobalConsts 用于封装常量信息；Java 工具类 HttpUtils 用于封装 Http 网络连接；Java 实体类 Train 用于封装火车时刻数据；Java 适配器类 TrainAdapter 用于封装 ListView 的适配器。

11.5.3 案例的实现步骤和思路

（1）创建 Android 项目。
（2）在 res 目录下的 layout 子目录中创建布局文件 activity_main.xml。
（3）在 res 目录下的 layout 子目录中创建布局文件 item.xml，用于 ListView 的条目布局。
（4）编写 GlobalConsts 文件，封装数据连接的常量信息。
（5）编写 HttpUtils 文件，封装网络连接的方法。
（6）编写 Train 文件，为实体类封装火车时刻的数据信息。
（7）编写 TrainAdapter 文件，继承 BaseAdapter，用于 ListView 的数据显示适配器。
（8）编写 MainActivity 文件，复写生命周期方法 onCreate。
① 在 onCreate 方法中，先调用父类的 onCreate 方法。
② 使用 setContentView 方法加载布局文件 activity_main.xml。
③ 获取界面按钮控件，并设置按钮监听器。在监听器中开启新线程，使用 HttpUtils 建立网络连接，获取 JSON 数据，再把 JSON 数据转换为 Train 实体，最后使用适配器 TrainAdapter，把数据显示在界面的 ListView 上。

11.5.4 案例参考代码

（1）布局文件 activity_main.xml 代码如下。

```xml
<LinearLayout xmlns:android="http://schemas.android.com/apk/res/android"
    xmlns:tools="http://schemas.android.com/tools"
    android:id="@+id/LinearLayout1"
    android:layout_width="match_parent"
    android:layout_height="match_parent"
    android:background="@drawable/background"
    android:orientation="vertical"
    android:paddingBottom="@dimen/activity_vertical_margin"
    android:paddingLeft="@dimen/activity_horizontal_margin"
    android:paddingRight="@dimen/activity_horizontal_margin"
    android:paddingTop="@dimen/activity_vertical_margin"
    tools:context=".MainActivity">

    <LinearLayout
        android:layout_width="fill_parent"
        android:layout_height="wrap_content"
        android:orientation="horizontal"
```

```xml
android:background="@android:color/white"
android:padding="10dp">

    <LinearLayout
        android:layout_width="wrap_content"
        android:layout_height="wrap_content"
        android:orientation="vertical"
        android:layout_gravity="center_vertical">

        <TextView
            android:layout_width="wrap_content"
            android:layout_height="wrap_content"
            android:text="出发地"
            android:textSize="23sp"/>

        <TextView
            android:layout_width="wrap_content"
            android:layout_height="wrap_content"
            android:text="目的地"
            android:textSize="23sp"
            android:layout_marginTop="10dp"/>

    </LinearLayout>

    <ImageView
        android:layout_width="wrap_content"
        android:layout_height="wrap_content"
        android:src="@drawable/iv01"
        android:layout_gravity="center_vertical"
        android:layout_marginLeft="5dp"
        android:layout_marginRight="5dp"/>

    <LinearLayout
        android:layout_width="wrap_content"
        android:layout_height="wrap_content"
        android:orientation="vertical">

        <EditText
            android:id="@+id/et_start"
            android:layout_width="match_parent"
            android:layout_height="wrap_content"
            android:background="@drawable/et_background"
            android:paddingLeft="10dp">
        </EditText>
```

```xml
        <EditText
            android:id="@+id/et_end"
            android:layout_width="match_parent"
            android:layout_height="wrap_content"
            android:background="@drawable/et_background"
            android:layout_marginTop="10dp"
            android:paddingLeft="10dp"/>

    </LinearLayout>

    <Button
        android:id="@+id/btn_submat"
        android:layout_width="match_parent"
        android:layout_height="wrap_content"
        android:background="@drawable/button_selector"
        android:text="查 询"
        android:textColor="@android:color/white"
        android:textSize="20sp"
        android:layout_marginTop="10dp"/>

    <ListView
        android:id="@+id/lv_train"
        android:layout_width="match_parent"
        android:layout_height="wrap_content"
        android:layout_marginTop="10dp">
    </ListView>

</LinearLayout>
```

（2）ListView 条目布局文件 item.xml 代码如下。

```xml
<?xml version="1.0" encoding="utf-8"?>
<LinearLayout xmlns:android="http://schemas.android.com/apk/res/android"
    android:layout_width="match_parent"
    android:layout_height="wrap_content"
    android:orientation="horizontal"
    android:gravity="center_vertical"
    android:padding="10dp">

    <LinearLayout
        android:layout_width="wrap_content"
        android:layout_height="wrap_content"
        android:orientation="vertical"
        android:gravity="center_horizontal">
```

```xml
<TextView
    android:id="@+id/tv_trainOpp"
    android:layout_width="wrap_content"
    android:layout_height="wrap_content"
    android:textSize="20sp"
    android:text="1462" />

<TextView
    android:id="@+id/tv_mileage"
    android:layout_width="wrap_content"
    android:layout_height="wrap_content"
    android:textSize="10sp"
    android:textColor="#6e6e6e"
    android:text="44481(km)" />

</LinearLayout>

<LinearLayout
    android:layout_width="wrap_content"
    android:layout_height="wrap_content"
    android:orientation="vertical"
    android:gravity="center_horizontal"
    android:layout_marginLeft="5dp">

    <ImageView
        android:id="@+id/imageView1"
        android:layout_width="wrap_content"
        android:layout_height="wrap_content"
        android:src="@drawable/start_station" />

    <TextView
        android:id="@+id/tv_leave_time"
        android:layout_width="wrap_content"
        android:layout_height="wrap_content"
        android:textSize="10sp"
        android:textColor="#fe9d00"
        android:text="19:19" />

</LinearLayout>

<TextView
    android:id="@+id/tv_start_station"
    android:layout_width="wrap_content"
```

```xml
        android:layout_height="wrap_content"
        android:text="上海虹桥" />

    <LinearLayout
        android:layout_width="wrap_content"
        android:layout_height="wrap_content"
        android:orientation="vertical"
        android:gravity="center_horizontal"
        android:layout_marginLeft="5dp"
        android:layout_marginRight="5dp">

        <TextView
            android:id="@+id/tv_train_typename"
            android:layout_width="wrap_content"
            android:layout_height="wrap_content"
            android:text="动车"
            android:textColor="#8b8b8b"
            android:textSize="10sp" />

        <ImageView
            android:id="@+id/imageView2"
            android:layout_width="30dp"
            android:layout_height="wrap_content"
            android:src="@drawable/arrows01" />

    </LinearLayout>

    <LinearLayout
        android:layout_width="wrap_content"
        android:layout_height="wrap_content"
        android:orientation="vertical"
        android:gravity="center_horizontal">

        <ImageView
            android:id="@+id/imageView3"
            android:layout_width="wrap_content"
            android:layout_height="wrap_content"
            android:src="@drawable/end_station" />

        <TextView
            android:id="@+id/tv_arrived_time"
            android:layout_width="wrap_content"
            android:layout_height="wrap_content"
            android:textSize="10sp"
```

```xml
        android:textColor="#00a5d5"
        android:text="21:03" />

    </LinearLayout>

    <TextView
        android:id="@+id/tv_end_station"
        android:layout_width="wrap_content"
        android:layout_height="wrap_content"
        android:text="苏州园区" />

    <LinearLayout
        android:layout_width="fill_parent"
        android:layout_height="wrap_content"
        android:orientation="vertical"
        android:gravity="right|center_vertical">

        <ImageView
            android:id="@+id/imageView4"
            android:layout_width="wrap_content"
            android:layout_height="wrap_content"
            android:src="@drawable/arrows02" />

    </LinearLayout>

</LinearLayout>
```

(3) Activity 组件 MainActivity 类代码如下。

```java
package com.example.android_demo11_2;

import java.io.IOException;
import java.util.ArrayList;
import java.util.List;
import org.apache.http.HttpEntity;
import org.apache.http.NameValuePair;
import org.apache.http.client.ClientProtocolException;
import org.apache.http.message.BasicNameValuePair;
import org.apache.http.util.EntityUtils;
import org.json.JSONArray;
import org.json.JSONException;
import org.json.JSONObject;
import android.app.Activity;
import android.os.Bundle;
import android.os.Handler;
```

```java
import android.os.Message;
import android.view.View;
import android.view.View.OnClickListener;
import android.widget.Button;
import android.widget.EditText;
import android.widget.ListView;
/**
 * 查询火车界面
 */
public class MainActivity extends Activity {
    //声明控件对象
    private EditText startET, endET;
    private Button submatBtn;
    private ListView lv;
    //声明适配器对象
    private TrainAdapter adapter;
    //声明 Handler 通信类
    private Handler handler=new Handler() {
        public void handleMessage(android.os.Message msg) {
            //更新列表方法
            adapter.updateData((ArrayList<Train>) msg.obj);
        };
    };
    /**
     * 初始化控件对象
     */
    private void setupView() {
        //获得控件对象
        startET=(EditText) this.findViewById(R.id.et_start);
        endET=(EditText) this.findViewById(R.id.et_end);
        submatBtn=(Button) this.findViewById(R.id.btn_submat);
        lv=(ListView) this.findViewById(R.id.lv_train);
        //声明 Adapter 适配器对象
        adapter=new TrainAdapter(this, null);
        //ListView 设置适配器对象
        lv.setAdapter(adapter);
    }

    /**
     * 添加监听器方法
     */
    private void addListener() {
        //单击查询按钮
```

```java
            submatBtn.setOnClickListener(new OnClickListener() {

                @Override
                public void onClick(View arg0) {
                    new Thread() {
                        public void run() {
            /*
             * 名称 类型 必填 说明 start string 是 出发站 end string 是 终点站
             * traintype string 否 列车类型,G-高速动车 K-快速 T-空调特快 D-动车组
             * Z-直达特快 Q-其他 key string 是 应用 APPKEY (应用详细页查
             询) dtype
             * string 否 返回数据的格式,xml 或 json, 默认 json
             */
                            List<NameValuePair>pairs=new ArrayList<NameValuePair>();
                            pairs.add(new BasicNameValuePair("start", startET
                                    .getText().toString().trim()));
                            pairs.add(new BasicNameValuePair("end", endET.getText()
                                    .toString().trim()));
                            pairs.add(new BasicNameValuePair("key",
                                    GlobalConsts.KEY));
                            try {
                                HttpEntity entity=HttpUtils.getEntity(
                                        GlobalConsts.URI, pairs);
                                String json=EntityUtils.toString(entity);
                                ArrayList<Train>trains=parseJson(json);
                                Message msg=Message.obtain(handler, 0, trains);
                                msg.sendToTarget();
                            } catch (ClientProtocolException e) {
                                //TODO Auto-generated catch block
                                e.printStackTrace();
                            } catch (IOException e) {
                                //TODO Auto-generated catch block
                                e.printStackTrace();
                            } catch (JSONException e) {
                                //TODO Auto-generated catch block
                                e.printStackTrace();
                            }
                        };
                    }.start();
                }
            });
        }

        /**
```

```java
 * 解析 Json 字符串
 * @param json
 * @return
 * @throws JSONException
 */
private ArrayList<Train>parseJson(String json) throws JSONException {
    //声明集合数据
    ArrayList<Train>trains=null;
    //声明 JSONObject 对象
    JSONObject object=new JSONObject(json);
    //获得反悔码
    String resultcode=object.optString("resultcode");
    //判断返回码是否成功
    if ("200".equals(resultcode)) {
        //获得 ArrayList 对象
        trains=new ArrayList<Train>();
        //数据解析
        String result=object.optString("result");
        JSONObject object2=new JSONObject(result);
        JSONArray array=object2.getJSONArray("data");
        for (int i=0; i <array.length(); i++) {
            JSONObject o1= (JSONObject) array.get(i);
            Train train=new Train();
            train.setTrainOpp(o1.optString("trainOpp"));
            train.setTrainTypeName(o1.optString("train_typename"));
            train.setStartStation(o1.optString("start_staion"));
            train.setEndStation(o1.optString("end_station"));
            train.setLeave_time(o1.optString("leave_time"));
            train.setArrived_time(o1.optString("arrived_time"));
            train.setMileage(o1.optString("mileage"));
            trains.add(train);
        }
    }
    return trains;
}

@Override
protected void onCreate(Bundle savedInstanceState) {
    super.onCreate(savedInstanceState);
    //设置布局
    setContentView(R.layout.activity_main);
    //初始化控件方法
    setupView();
    //添加监听器方法
```

```
        addListener();
    }
}
```

(4) GlobalConsts 类代码如下。

```
package com.example.android_demo11_2;

public class GlobalConsts {
    //聚合数据接口
    public static final String URI="http://apis.juhe.cn/train/s2s";
    //接口使用的 keyID
    public static final String KEY="6a92888ed0709c637650a64ebb0fbb3c";
}
```

(5) HttpUtils 类代码如下。

```
package com.example.android_demo11_2;

import java.io.IOException;
import java.util.List;
import org.apache.http.HttpEntity;
import org.apache.http.HttpResponse;
import org.apache.http.HttpStatus;
import org.apache.http.NameValuePair;
import org.apache.http.client.ClientProtocolException;
import org.apache.http.client.HttpClient;
import org.apache.http.client.methods.HttpGet;
import org.apache.http.client.methods.HttpUriRequest;
import org.apache.http.impl.client.DefaultHttpClient;
import org.apache.http.params.CoreConnectionPNames;

public class HttpUtils {
    public static final int METHOD_GET=0;
    public static final int METHOD_POST=1;

    /**
     * 请求实体
     *
     * @param uri
     * @param pairs
     * @return HttpEntity
     * @throws IOException
     * @throws ClientProtocolException
     */
    public static HttpEntity getEntity(String uri, List<NameValuePair>pairs)
            throws ClientProtocolException, IOException {
```

```java
        HttpEntity entity=null;
        //创建客户端
        HttpClient client=new DefaultHttpClient();
        //设置访问服务器超时
        client.getParams().setParameter(
                CoreConnectionPNames.CONNECTION_TIMEOUT, 3000);
        StringBuilder sb=new StringBuilder(uri);
        if (pairs !=null && !pairs.isEmpty()) {
            sb.append('?');
            for (NameValuePair pair : pairs) {
                sb.append(pair.getName()).append('=').append(pair.getValue())
                        .append('&');
            }
            sb.deleteCharAt(sb.length() -1);
        }
        //创建请求
        HttpUriRequest request=new HttpGet(sb.toString());
        //执行请求获得响应
        HttpResponse response=client.execute(request);
        //判断响应码
        if (response.getStatusLine().getStatusCode()==HttpStatus.SC_OK) {
            entity=response.getEntity();
        }
        //解析响应对象
        return entity;
    }
}
```

(6) Train 类代码如下。

```java
package com.example.android_demo11_2;
/**
 * 火车实体类
 */
public class Train {

    private String trainOpp;
    private String trainTypeName;
    private String startStation;
    private String endStation;
    private String leave_time;
    private String arrived_time;
    private String mileage;
    public String getTrainOpp() {
        return trainOpp;
    }
```

```java
    public void setTrainOpp(String trainOpp) {
        this.trainOpp=trainOpp;
    }
    public String getTrainTypeName() {
        return trainTypeName;
    }
    public void setTrainTypeName(String trainTypeName) {
        this.trainTypeName=trainTypeName;
    }
    public String getStartStation() {
        return startStation;
    }
    public void setStartStation(String startStation) {
        this.startStation=startStation;
    }
    public String getEndStation() {
        return endStation;
    }
    public void setEndStation(String endStation) {
        this.endStation=endStation;
    }
    public String getLeave_time() {
        return leave_time;
    }
    public void setLeave_time(String leave_time) {
        this.leave_time=leave_time;
    }
    public String getArrived_time() {
        return arrived_time;
    }
    public void setArrived_time(String arrived_time) {
        this.arrived_time=arrived_time;
    }
    public String getMileage() {
        return mileage;
    }
    public void setMileage(String mileage) {
        this.mileage=mileage;
    }
    public Train(String trainOpp, String trainTypeName, String startStation,
            String endStation, String leave_time, String arrived_time,
            String mileage) {
        super();
        this.trainOpp=trainOpp;
        this.trainTypeName=trainTypeName;
        this.startStation=startStation;
```

```java
            this.endStation=endStation;
            this.leave_time=leave_time;
            this.arrived_time=arrived_time;
            this.mileage=mileage;
        }
        public Train() {
            super();
        }
        @Override
        public String toString() {
            return "Train [trainOpp="+trainOpp+", trainTypeName="
                    +trainTypeName+", startStation="+startStation
                    +", endStation="+endStation+", leave_time="+leave_time
                    +", arrived_time="+arrived_time+", mileage="+mileage
                    +"]";
        }


}
```

（7）TrainAdapter 类代码如下。

```java
package com.example.android_demo11_2;

import java.util.ArrayList;

import android.content.Context;
import android.view.LayoutInflater;
import android.view.View;
import android.view.ViewGroup;
import android.widget.BaseAdapter;
import android.widget.TextView;
/**
 * 适配器类
 */
public class TrainAdapter extends BaseAdapter {

    //布局加载器
    private LayoutInflater inflater;
    //数据集合
    private ArrayList<Train>trains;

    /**
     * 构造方法
     * @param context
     * @param trains
```

```java
     */
    public TrainAdapter(Context context, ArrayList<Train>trains) {
        //判断数据是否为NULL
        if (trains==null) {
            this.trains=new ArrayList<Train>();
        } else {
            this.trains=trains;
        }
        //获得布局加载起对象
        this.inflater=LayoutInflater.from(context);
    }
    /**
     * 更新数据方法
     * @param trains
     */
    public void updateData(ArrayList<Train>trains) {
        if (trains==null) {
            this.trains=new ArrayList<Train>();
        } else {
            this.trains=trains;
        }
        //更新数据列表
        this.notifyDataSetChanged();
    }

    @Override
    public int getCount() {
        //TODO Auto-generated method stub
        return trains.size();
    }

    @Override
    public Object getItem(int position) {
        //TODO Auto-generated method stub
        return trains.get(position);
    }

    @Override
    public long getItemId(int arg0) {
        //TODO Auto-generated method stub
        return 0;
    }

    @Override
    public View getView(int position, View contentView, ViewGroup arg2) {
```

```java
//声明换成类对象
ViewHolder holder=null;
//判断控件是否为Null
if(contentView==null){
    contentView=this.inflater.inflate(R.layout.item, null);
    //声明缓存类对象
    holder=new ViewHolder();
    //获得布局控件
    holder.trainOppTv= (TextView)contentView
        .findViewById(R.id.tv_trainOpp);
    holder.trainTypeNameTv= (TextView)contentView
        .findViewById(R.id.tv_train_typename);
    holder.startStationTv= (TextView)contentView
        .findViewById(R.id.tv_start_station);
    holder.endStationTv= (TextView)contentView
        .findViewById(R.id.tv_end_station);
    holder.leaveTimeTv= (TextView)contentView
        .findViewById(R.id.tv_leave_time);
    holder.arrivedTimeTv= (TextView)contentView
        .findViewById(R.id.tv_arrived_time);
    holder.mileageTv= (TextView)contentView
        .findViewById(R.id.tv_mileage);
    //缓冲类对象存放view控件中
    contentView.setTag(holder);
}else{
    //获得存储的控件
    holder= (ViewHolder)contentView.getTag();
}
//获得火车实体对象
Train train=trains.get(position);
//判断火车对象是否为NULL
if(train !=null){
    //设置布局
    if(train.getTrainOpp().length()>=4 )
        holder.trainOppTv
            .setText(train.getTrainOpp().substring(0, 4)+"...");
    else
        holder.trainOppTv.setText(train.getTrainOpp());
    holder.trainTypeNameTv.setText(train.getTrainTypeName());
    holder.startStationTv.setText(train.getStartStation());
    holder.endStationTv.setText(train.getEndStation());
    holder.leaveTimeTv.setText(train.getLeave_time());
    holder.arrivedTimeTv.setText(train.getArrived_time());
    holder.mileageTv.setText(train.getMileage()+"(公里)");
}
//返回控件
```

```
        return contentView;
    }
    /**
     * 缓冲类
     */
    class ViewHolder {
        TextView trainOppTv, trainTypeNameTv, startStationTv, endStationTv,
                leaveTimeTv, arrivedTimeTv, mileageTv;
    }
}
```

上述代码中通过 new DefaultHttpClient()方法获得 HttpClient 对象,并设置访问的超时时间,最后通过 HttpClient 对象的 execute()方法发送请求,之后通过返回的 HttpResponse 对象中的 getEntity()方法获得服务器返回的相应数据。

11.5.5 案例运行效果

案例运行效果如图 11-2 所示,当用户在两个文本框中分别输入"出发地"和"目的地"之后,单击"查询"按钮,如果查询到结果,会在下方的 ListView 中显示出来。

图 11-2 案例 HttpClient 接口运行效果

习 题 11

1. 什么是 URI?
2. 使用 HttpClient 接口的一般步骤是什么?

第 12 章

Android 管理器

本章主要为各位读者介绍 Android 的两个常用管理器,分别是 TelephonyManager(电话管理器)和 SmsManager(短信管理器),通过这两个管理器在开发中就可以非常方便地发送短信或者获得手机的通话状态信息,如手机通话中的状态、挂断电话的状态等。

12.1 电话管理器

TelephonyManager 是一个管理手机通话状态、电话网络信息的服务。
在程序中获取 TelephonyManager 十分简单,只要调用如下代码即可。

```
TelephonyManager manager=
    (TelephonyManager)this.getSystemService(TELEPHONY_SERVICE);
```

下面将通过一个案例来说明 Android 中电话管理器的基本使用方法。

12.2 案例 TelephonyManager

12.2.1 案例功能描述

案例中界面有一个 TextView 控件,该案例主要通过讲解监听手机通话状态,来帮助读者了解和使用电话管理器是如何使用、并如何捕捉电话状态的。

12.2.2 案例程序结构

案例中包括一个布局文件(activity_main.xml),一个 Activity 组件类(MainActivity 类),用于实现用户界面交互功能。

12.2.3 案例的实现步骤和思路

(1) 创建 Android 项目。
(2) 在 res 目录下的 layout 子目录中创建一个新的布局文件 activity_main.xml。

(3) 编写 MainActivity 文件，复写生命周期方法 onCreate。

① 在 onCreate 方法中，先调用父类的 onCreate 方法；
② 使用 setContentView 方法加载布局文件 activity_main.xml；
③ 获取界面 Textview 控件，把监控到的手机状态显示在界面上；
④ 创建内部类，继承 PhoneStateListener，来监听手机电话的状态。

12.2.4 案例参考代码

（1）布局文件 activity_main.xml 的代码如下。

```xml
<?xml version="1.0" encoding="utf-8"?>
<LinearLayout xmlns:android="http://schemas.android.com/apk/res/android"
android:orientation="vertical"
android:layout_width="fill_parent"
android:layout_height="fill_parent">

    <TextView
        android:id="@+id/myTextView1"
        android:layout_width="fill_parent"
        android:layout_height="wrap_content"
        android:text="@string/hello_world"
        android:textSize="20sp">
    </TextView>

</LinearLayout>
```

（2）Activity 组件 MainActivity 类代码如下。

```java
package com.example.android_demo12_1;

import android.app.Activity;
import android.content.ContentResolver;
import android.content.Context;
import android.database.Cursor;
import android.graphics.Color;
import android.os.Bundle;
import android.provider.ContactsContract.CommonDataKinds.Phone;
import android.telephony.PhoneStateListener;
import android.telephony.TelephonyManager;
import android.widget.TextView;
/**
 * 电话管理器
 */
public class MainActivity extends Activity {
```

```java
    private TextView myTextView1;

    @Override
    public void onCreate(Bundle savedInstanceState) {
        super.onCreate(savedInstanceState);
        setContentView(R.layout.activity_main);
        myTextView1=(TextView) findViewById(R.id.myTextView1);
        /*新增的 PhoneStateListener*/
        MyPhoneCallListener myPhoneCallListener=new MyPhoneCallListener();
        /*取得电话服务*/
        TelephonyManager tm=(TelephonyManager) this
                .getSystemService(Context.TELEPHONY_SERVICE);
        /*注册 Listener*/
        tm.listen(myPhoneCallListener, PhoneStateListener.LISTEN_CALL_STATE );
    }

    /*内部 class 继承 PhoneStateListener*/
    public class MyPhoneCallListener extends PhoneStateListener {
        /*重写 onCallStateChanged,当状态改变时改变 myTextView1 的文字及颜色*/
        public void onCallStateChanged(int state, String incomingNumber) {
            switch (state){
            /*无任务状态时*/
            case TelephonyManager.CALL_STATE_IDLE:
                myTextView1.setTextColor(getResources()
                        .getColor(R.drawable.red));
                myTextView1.setText("CALL_STATE_IDLE");
                break;
            /*接起电话时*/
            case TelephonyManager.CALL_STATE_OFFHOOK:
                myTextView1.setTextColor(getResources().getColor(
                        R.drawable.green));
                myTextView1.setText("CALL_STATE_OFFHOOK");
                break;
            /*电话进来时*/
            case TelephonyManager.CALL_STATE_RINGING:
                getContactPeople(incomingNumber);
                break;
            default:
                break;
            }
            super.onCallStateChanged(state, incomingNumber);
        }
    }
}
```

```
private void getContactPeople(String incomingNumber) {
    myTextView1.setTextColor(Color.BLUE);
    ContentResolver contentResolver=getContentResolver();
    Cursor cursor=null;
    /*cursor 里要放的字段名称*/
    String[] projection=new String[] { Phone._ID, Phone.DISPLAY_NAME,
        Phone.NUMBER };
    /*用来电电话号码查找该联系人*/
    cursor=contentResolver.query(Phone.CONTENT_URI, projection,
        Phone.NUMBER+"=?", new String[] { incomingNumber }, "");
    /*找不到联系人*/
    if (cursor.getCount()==0) {
        myTextView1.setText("unknown Number:"+incomingNumber);
    } else if (cursor.getCount()>0) {
        cursor.moveToFirst();
        /*获取联系人的姓名和电话号码*/
        String name=cursor.getString(cursor
            .getColumnIndex(Phone.DISPLAY_NAME));
        String number=cursor.getString(cursor
            .getColumnIndex(Phone.NUMBER));
        myTextView1.setText(name+":"+number);
    }
}
```

上述代码中,PhoneStateListener 提供了监听电话状态等事件的方法,所以,要监控手机电话状态,需要创建 PhoneStateListener 对象,并重写其中的 onCallStateChanged() 方法。通过传入的 state 判断电话状态,上述代码中只针对 CALL_STATE_IDLE、CALL_STATE_OFFHOOK 及 CALL_STATE_RINGING 这三种状态,而这三种状态分别代表"待机"、"通话中"和"来电响铃中"。

在代码中,通过 getSystemService 获取 TelephonyManager 对象,向 TelephonyManager 注册 PhoneCallListener。注册时,传入 PhoneCallListener 类及要监听的事件名称,因监听的是电话状态,注册中需要传递 LISTEN_CALL_STATE。

12.2.5 案例运行效果

案例运行效果如图 12-1 所示。

需要指出的是,由于该程序需要获取手机的通话状态,因此必须在 AndroidManifest.xml 文件中增加如下权限配置代码:

```
<uses-permission
    android:name="android.permission.READ_PHONE_STATE"/>
```

图 12-1　监听来电状态

12.3　短信管理器

　　SmsManager 是 Android 提供的另一个非常常见的服务，SmsManager 提供了一系列 sendXXXMessage() 方法用于发送短信，不过就现在的实际应用来看，短信通常都是普通的文本内容，也就是调用 sendTextMessage() 方法进行发送即可。

　　下面将通过 SmsManager 案例说明 Android 中短信管理器的基本使用方法。

12.4　案例 SmsManager

12.4.1　案例功能描述

　　案例中界面上有两个 EditText 控件和一个按钮，单击按钮后会给指定手机发送短信。

12.4.2　案例程序结构

　　案例中包括一个布局文件（activity_main.xml），一个 Activity 组件类（MainActivity 类），用于实现用户界面交互功能。

12.4.3　案例的实现步骤和思路

　　（1）创建 Android 项目。
　　（2）在 res 目录下的 layout 子目录中创建一个新的布局文件 activity_main.xml。

(3) 编写 MainActivity 文件,复写生命周期方法 onCreate。
① 在 onCreate 方法中,先调用父类的 onCreate 方法;
② 使用 setContentView 方法加载布局文件 activity_main.xml;
③ 获取界面的三个控件,给 Button 设置监听器;
④ 在按钮监听器的 onClick 方法中获取输入框内容,并使用短信管理器发送短信。

12.4.4 案例参考代码

(1) 布局文件 activity_main.xml 代码如下。

```xml
<LinearLayout xmlns:android="http://schemas.android.com/apk/res/android"
    xmlns:tools="http://schemas.android.com/tools"
    android:id="@+id/LinearLayout1"
    android:layout_width="match_parent"
    android:layout_height="match_parent"
    android:orientation="vertical"
    android:paddingBottom="@dimen/activity_vertical_margin"
    android:paddingLeft="@dimen/activity_horizontal_margin"
    android:paddingRight="@dimen/activity_horizontal_margin"
    android:paddingTop="@dimen/activity_vertical_margin"
    tools:context=".MainActivity">

    <EditText
        android:id="@+id/et_phonenumber"
        android:layout_width="match_parent"
        android:layout_height="wrap_content"
        android:ems="10"
        android:inputType="phone"
        android:hint="请输入手机号">
        <requestFocus />
    </EditText>

    <EditText
        android:id="@+id/et_content"
        android:layout_width="match_parent"
        android:layout_height="wrap_content"
        android:layout_marginTop="10dp"
        android:hint="请输入短信内容"
        android:ems="10" />

    <Button
        android:id="@+id/btn_sendMessage"
        android:layout_width="match_parent"
        android:layout_height="wrap_content"
```

```xml
            android:layout_marginTop="10dp"
            android:text="发送短信" />

</LinearLayout>
```

(2) Activity 组件 MainActivity 类代码如下。

```java
package com.example.android_demo12_2;

import android.app.Activity;
import android.os.Bundle;
import android.telephony.SmsManager;
import android.view.View;
import android.view.View.OnClickListener;
import android.widget.Button;
import android.widget.EditText;
/**
 * 发送短信
 */
public class MainActivity extends Activity {
    //声明编辑文本框对象
    private EditText phoneNumber, content;
    //发送短信按钮
    private Button sendMessage;
    //短信管理器
    private SmsManager sManager;

    @Override
    protected void onCreate(Bundle savedInstanceState) {
        super.onCreate(savedInstanceState);
        //设置布局
        setContentView(R.layout.activity_main);
        //获得控件对象
        phoneNumber=(EditText) this.findViewById(R.id.et_phonenumber);
        content=(EditText) this.findViewById(R.id.et_content);
        sendMessage=(Button) this.findViewById(R.id.btn_sendMessage);
        //获得短信管理类对象
        sManager=SmsManager.getDefault();
        //发送按钮
        sendMessage.setOnClickListener(new OnClickListener() {

            @Override
            public void onClick(View arg0) {
                //发送短信
                sManager.sendTextMessage(phoneNumber.getText().toString(),
```

```
                null, content.getText().toString(), null, null);
        }
    });
}
```

12.4.5 案例运行效果

案例运行效果如图 12-2 所示。

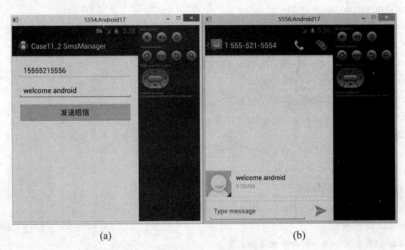

图 12-2 发送短信

习 题 12

1. 如何通过 TelephonyManager 获得电话的运行状态？
2. 简述短信管理器的作用及其使用方法。

第 13 章

LBS 定位服务

13.1 LBS 简介

LBS(Location Based Service)被称作基于位置的服务,是通过移动通信运营商提供的无线通信网络或 GPS(Global Positioning System,全球定位系统)定位方式获取移动设备用户位置信息的一种业务。运营商或应用软件可以通过该业务为用户提供相应的增值服务。LBS 的应用在近两年发展迅速,时下很多流行的手机应用软件,如百度地图、滴滴打车、美团等,都采用了 LBS 位置服务。

传统意义上讲,LBS 包括两层含义:定位和服务。LBS 首先是确定移动设备所在的地理位置,例如用户外出前往某地,人生地不熟,想要知道自己当前所在的位置信息,可以使用定位软件,如 Google 地图、百度地图等。定位服务需要在互联网或无线网络的环境中进行,并且使用到地理信息系统(Geographic Information System,GIS)。如果用户想要住宿或者就餐,可以使用提供与位置相关的各类信息服务软件,查找附近的酒店、餐馆信息。

LBS 的概念虽然提出的时间不长,但其发展已经有相当长的一段历史。LBS 起源于美国以军事应用为目的所部署的 GPS,随后在测绘和车辆跟踪定位等领域开始应用。当 GPS 民用化以后,产生了以定位为核心功能的大量应用,直到 20 世纪 90 年代后期,LBS 及其所涉及的技术才得到广泛的重视和应用。

我国的 LBS 商业应用始于 2001 年中国移动首次开通的移动梦网品牌下的位置服务。2003 年,中国联通又推出了"定位之星"业务。用户在使用这项服务时,只要在手机上输入出发地和目的地,就可以查到开车路线;如果用语音导航,还能得到实时提示,该项业务还能够实现 5~50 m 的连续、精确定位,用户可以在较快的速度下体验下载地图和导航类的复杂服务。但是由于当时移动通信的带宽很窄、GPS 的普及率比较低,最重要的是市场需求并不旺盛,所以,几家规模较大的运营商虽然热情很高,但是整个市场并没有像预期的那样顺利启动。在 2001 年之后,LBS 在中国发展的近 4 年时间里始终处于非常缓慢的增长阶段。

2006 年初,中国移动在北京、天津、辽宁、湖北 4 个省市进行了"手机地图"业务的试点运行,为广大手机用户提供显示、动态缩放、动态漫游跳转、全图、索引图、比例尺、城市切换以及各种查询等位置服务。互联网地图的出现加速了我国 LBS 产业的发展。众多地图厂商、软件厂商相继开发了一系列在线的 LBS 终端软件产品。此后,随着中国 3G 网络部署提速 GPS 手机的大力推广,LBS 行业在国内迎来一个爆发增长期。

2011 年以后,Android 占领主流移动互联网开发市场,使得基于 Android 平台的 LBS 有了更广阔的发展空间和巨大的市场需求。从导航和团购应用到社交软件,Android LBS 的发展前景被广泛看好。有理由相信,Android 平台以其开放的特性,加之广大开发者的努力,终将使 LBS 产业发展进入一个全新的局面。

13.2 LBS 服务模式

当前市场下随着用户对多元化 LBS 服务需求不断增长,出现了以下几种新兴的服务模式,下面将逐一进行介绍。

13.2.1 社交网络和游戏模式

该模式起源于 2009 年在美国上线的 Foursquare,是一个专注于 LBS 的网站。该网站的商业模式俗称"切客"(Check-in)模式,需要用户以主动签到的游戏方式,记录和分享自己所在的当前地理位置信息,帮助用户与外部世界创建更加广泛和密切的联系。同时通过与商家合作,对用户提供优惠或折扣的奖励,可以很好地为商户或品牌进行各种形式的营销和推广。国内基于该模式的应用有切客、街旁、嘀咕等。然而随着移动互联网的发展,签到模式的社交应用大多受困于盈利模式的局限慢慢被淘汰,取而代之的是以微信、微博、陌陌等更为强势的、具有更广泛用户基础的社交应用。这些软件由于其内容特性和盈利模式往往对用户更具黏性,在它们身上所嵌入的 LBS 功能,比单纯的签到软件要更具市场价值。例如微信中的 LBS 应用如图 13-1 所示。

图 13-1 微信 LBS 应用

该软件使用 LBS 定位所有用户的位置信息,方便用户查找附近的社交对象。

以 MyTown 和 16Fun 为代表的现实大富翁类基于 LBS 的游戏则有所不同。这种应用以游戏为主、社交为辅,可以让用户利用手机购买现实地理位置里的虚拟房产与道具,并进行消费与互动。该模式将现实和虚拟真正进行融合,更具趣味性,可玩性与互动性更强,比 Check-in 模式更具黏性。盈利模式则以联合商家营销和提供增值服务为主,同时也会植入广告。

13.2.2 生活信息服务模式

以点评网或者生活信息类网站与地理位置服务结合的模式，为用户提供搜索、评论和分享本地餐饮、休闲、娱乐等生活信息。该模式借助 LBS 实现生活服务信息的增值，在为用户提供更客观、更准确的生活信息指南的同时，也为商户提供了一个基于位置的精准营销平台。

该模式在现阶段的代表应用是以大众点评、美团为首的各类团购软件。近两年该类应用在国内市场迅速崛起并竞争激烈，甚至于一些大型互联网公司，如百度、阿里巴巴等都参与其中，其服务领域也逐渐渗透到旅游、票务、信用卡等日常生活服务的方方面面。美团网中的 LBS 应用如图 13-2 所示。

该类应用盈利模式主要是为线下实体生活服务商户有偿提供关键字搜索、电子优惠券、品牌推广、互动营销等多种互联网推广服务。

13.2.3 电子商务模式

在激烈的市场竞争下，一些企业开始尝试 LBS 和电子商务的合作模式。最早的是团购网站糯米网和美国的 GroupTabs，尝试将优惠券嫁接 LBS 签到技术实现新的社交化电子商务模式。用户通过应用签到后可以得到实体商户的折扣和优惠券，再去实体商户消费时便可享受折扣或优惠。当下最流行的电子商务模式 LBS 应用当属滴滴打车。滴滴打车中的 LBS 应用如图 13-3 所示。

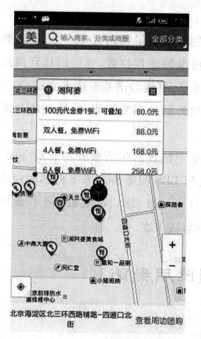

图 13-2　美团 LBS 应用

图 13-3　滴滴打车 LBS 应用

该应用是为用户提供可以抵付现金的打车券,并提供网上支付模式,旨在吸引更多的用户打造自己的商业平台。

可以看到,LBS 的服务模式正在趋于多元化、普遍化。尤其在现阶段的各类 Android 应用中,LBS 的嵌入已经成为一项标准配置。

13.3 获取位置信息

Android 定位方式一般有两种：GPS 和网络提供者 NLP(Android Network Location Provider)。GPS 定位的优点是定位信息比较精确,平均精度在 10m 左右;缺点则是信息返回较慢,定位时间往往在几十秒到几分钟不等,且只能在户外使用,耗电严重。NLP 通过基站和 WiFi 信号获取位置信息,优点是室内外均可使用、耗电少、定位时间短,一般只需几秒;缺点则是定位不够精确。

现阶段的 Android LBS 开发通常使用第三方提供的地图应用接口。这些地图应用可以通过 SDK 嵌入到程序中,常用的有 Google 地图、百度地图和高德地图等。

Google 地图是 Google 公司 2005 年推出的一款电子地图服务,包括局部详细的卫星照片。此款服务可以提供含有政区和交通以及商业信息的矢量地图、不同分辨率的卫星照片和可以用来显示地形和等高线地形视图。Google 地图是开发国际地图服务的最佳选择,但由于 2014 年 5 月 27 日之后 Google 的相关网站开始无法在中国地区访问,想要更新并使用 Google 地图的 SDK 也变得非常困难。

百度地图是百度提供的一项网络地图搜索服务,覆盖了国内近 400 个城市、数千个区县。在百度地图里,用户可以查询街道、商场、楼盘的地理位置,也可以找到离您最近的所有餐馆、学校、银行、公园等。2014 年 12 月 15 日,百度与诺基亚达成协议,未来诺基亚地图及导航业务 Here 将向百度提供中国内地以外的地图数据服务,这意味着不久的将来开发者可以使用百度地图开发国际地图服务。

高德地图是国内一流的地图导航产品,也是基于位置的生活服务功能最全面、信息最丰富的手机地图,由国内最大的电子地图、导航和 LBS 服务解决方案提供商高德软件提供。其数据覆盖中国大陆及香港、澳门,遍及 337 个地级、2857 个县级以上行政区划单位。高德 LBS 平台可以为开发者提供免费的地图解决方案。

13.4 百度地图使用案例

13.4.1 案例概述

本案例以使用百度地图 SDK 开发 Android 平台的定位服务应用为例,带领读者了解 LBS 第三方 SDK 的使用流程及开发方式,熟练掌握 LBS 应用开发所需的知识要点和基本技术。百度地图应用的界面效果如图 13-4 所示。

13.4.2 案例分析

百度地图 Android SDK 是一套基于 Android 2.1 及以上版本设备的应用程序接口。开发者可以使用该 SDK 开发适用于 Android 系统移动设备的地图应用，通过调用地图 SDK 接口，可以轻松访问百度地图服务和数据，构建功能丰富、交互性强的地图类应用程序。百度地图 Android SDK 提供的所有服务都是免费的，接口使用无次数限制，但需要申请密钥（key）后，才可使用百度地图 Android SDK。

百度地图 Android SDK 具有 10 项特色功能，其官方介绍如下。

（1）地图：提供地图展示和操作功能。

（2）POI 检索：支持周边检索、区域检索和城市内检索。

（3）地理编码：提供地理坐标和地址之间的相互转换。

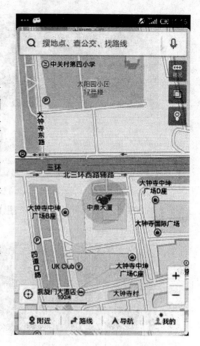

图 13-4　百度地图

（4）线路规划：支持公交信息查询、公交换乘查询、驾车线路规划和步行路径检索。

（5）地图覆盖物：支持多种地图覆盖物，帮助开发者展示更丰富的地图。

（6）定位：采用 GPS、WiFi、基站、IP 混合定位模式。

（7）离线地图：使用离线地图可节省用户流量，提供更好的地图展示效果。

（8）导航：支持调启百度地图客户端导航和调启 Web 页面导航。

（9）LBS 云：针对 LBS 开发者全新推出的平台级服务，使用其可以实现移动设备开发者存储海量位置数据的服务器，且支持高效检索用户数据。

（10）特色功能：包括短串分享、Place 详情信息检索、热力图等。

可以看到，该 SDK 基本覆盖了 LBS 开发中所用到的所有功能。可以从百度地图 API 官网下载并使用该 SDK。需要注意的是开发 LBS 应用需要在移动设备环境下测试，而不能使用虚拟机 Android 模拟器。

13.4.3 案例实现

本案例的实现大致需要以下几个步骤，依次进行讲解。

1. 下载 SDK

首先登录百度地图 API 官网，如图 13-5 所示。

在 Android 开发中选择下载 SDK，如图 13-6 所示。

单击后进入下载页面，如图 13-7 所示。

图 13-5 百度地图 API 官网

图 13-6 下载 SDK

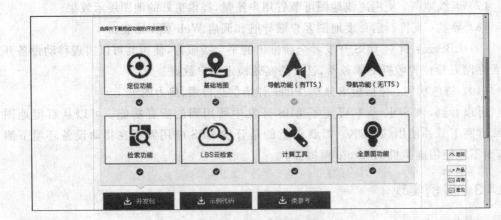

图 13-7 SDK 下载界面

在这里用户可以依据自己要开发的应用功能选择下载。对于每一个功能,百度都提供了开发包 SDK、示例代码和参考类供开发者使用。本例中选择全部 7 项功能并单击下载开发包,下载后的压缩包如图 13-8 所示。

将 SDK 压缩包解压并将 lib 文件夹覆盖到项目 JAR 包下,如图 13-9 所示。

第13章 LBS定位服务

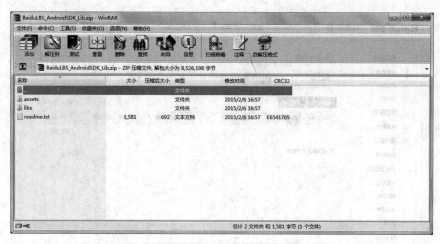

图13-8　SDK压缩包

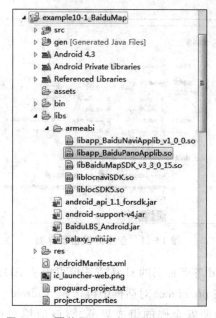

图13-9　覆盖导入SDK后的项目JAR包

2. 申请开发密钥

要在程序中使用各种LBS功能，需要首先申请开发密钥。首先注册一个百度账号（如果没有的情况下），或使用第三方账号登录。登录后的界面如图13-10所示。

如图13-10所示，单击创建应用按钮，弹出创建应用界面。在该界面下可以为应用申请开发密钥。写入应用名称并在应用类型的下拉列表中选择Android SDK，选择将要启用的服务，如图13-11所示。

安全码的组成规则为：Android签名证书的sha1值＋;＋packagename（即数字签

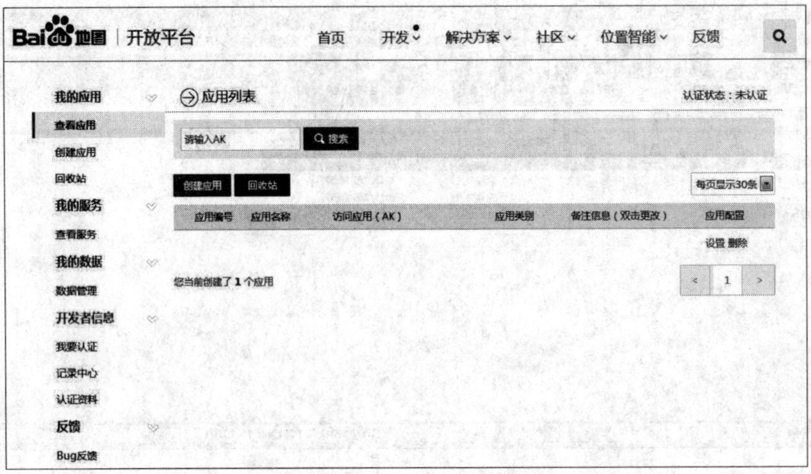

图 13-10　申请应用密钥

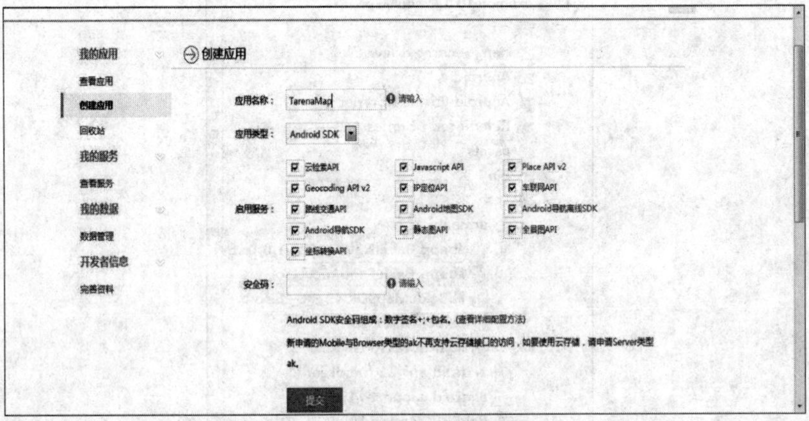

图 13-11　创建开发密钥界面

名+分号+包名）。Android 签名证书的 sha1 密钥值可以从 ADT 中直接查看，选择菜单项 Windows->Preference，在 preference 对话框中再选择 Android->Build，如图 13-12 所示。

查看 sha1 值，之后参照安全码的组成规范，将该值填写在安全码一栏中，单击"提交"按钮，则新建的应用开发密钥如图 13-13 所示。本例中的安全码为 71:EF:05:AB:73:92:8B:13:A4:2D:FE:97:9B:1D:55:5A:E9:3C:0F:79;com.example.example10_1_baidumap。

该密钥是独立且唯一的，使用该密钥可以得到百度地图 SDK 相关的访问权限，用以完成 LBS 应用的开发工作。

3. 开发 LBS 应用

要在应用中使用百度地图数据的接口，首先需要将申请到的密钥添加到 AndroidManifest 注册信息中，并配置相关权限。在 Application 标签下加入密钥的代码如下所示。

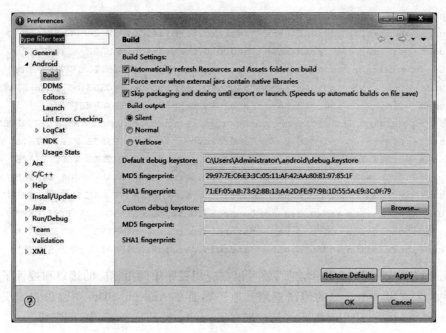

图 13-12 查看 sha1 密钥值

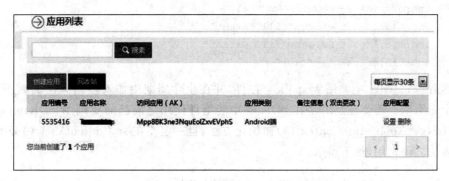

图 13-13 创建好的开发密钥

```
<application
    android:allowBackup="true"
    android:icon="@drawable/ic_launcher"
    android:label="@string/app_name"
    android:theme="@style/AppTheme">
    <meta-data
        android:name="com.baidu.lbsapi.API_KEY"
        android:value="ukSWG04LXvLw9IlNSsTseHL7" />
...
</application>
```

如上所示,密钥名称不变,需要将密钥值 value 设置为之前自己申请的开发密钥。然后加入所需的应用权限,如下所示。

```xml
<uses-permission android:name="android.permission.GET_ACCOUNTS" />
<uses-permission android:name="android.permission.USE_CREDENTIALS" />
<uses-permission android:name="android.permission.MANAGE_ACCOUNTS" />
<uses-permission android:name="android.permission.AUTHENTICATE_ACCOUNTS" />
<uses-permission android:name="android.permission.ACCESS_NETWORK_STATE" />
<uses-permission android:name="android.permission.INTERNET" />
<uses-permission android:name="com.android.launcher.permission.READ_SETTINGS" />
<uses-permission android:name="android.permission.CHANGE_WIFI_STATE" />
<uses-permission android:name="android.permission.ACCESS_WIFI_STATE" />
<uses-permission android:name="android.permission.READ_PHONE_STATE" />
<uses-permission android:name="android.permission.WRITE_EXTERNAL_STORAGE" />
<uses-permission android:name="android.permission.BROADCAST_STICKY" />
<uses-permission android:name="android.permission.WRITE_SETTINGS" />
<uses-permission android:name="android.permission.READ_PHONE_STATE" />
```

配置好 AndroidManifest 之后,就可以在应用程序中调用相应的接口和控件了。

我们先来做一个简单的地图应用。地图界面在 Android SDK 中是以自定义 View 控件的形式提供的,使用时需要将其布局在 Activity 或 Fragment 中,如下所示。

```xml
<com.baidu.mapapi.map.MapView
    android:id="@+id/bmapView"
    android:layout_width="fill_parent"
    android:layout_height="fill_parent"
    android:clickable="true" />
```

然后在 activity 中调用 MapView 控件,并在生命周期中加入控制方法的调用,管理地图生命周期。需要注意的是在 SDK 各功能组件使用之前都需要调用 SDKInitializer.initialize(getApplicationContext())初始化方法,且一定要在 setContentView()等组件初始化方法之前调用,如下所示。

```java
public class MainActivity extends Activity {
    MapView mMapView=null;
    @Override
    protected void onCreate(Bundle savedInstanceState) {
        super.onCreate(savedInstanceState);
        SDKInitializer.initialize(getApplicationContext());
        setContentView(R.layout.activity_main);
        mMapView= (MapView) findViewById(R.id.bmapView);
    }
    @Override
    protected void onResume() {
        super.onResume();
        mMapView.onResume();
    }
    @Override
```

```
protected void onPause() {
    super.onPause();
    mMapView.onPause();
}
@Override
protected void onDestroy() {
    super.onDestroy();
    mMapView.onDestroy();
}
}
```

运行该案例,效果如图 13-14 所示。

实现了地图功能之后,可以在其基础上实现更多其他的 LBS 功能。例如现在要实现一个定位的功能,首先需要在 AndroidManifest 中静态注册一个用于处理远程定位信息的后台服务。该服务在百度地图定位 API 架包中,如下所示。

```
<service
    android:name="com.baidu.location.f"
    android:enabled="true"
    android:process=":remote">
</service>
```

图 13-14 基本地图功能

注册了该服务的应用就可以接收定位信息了。声明定位相关的全局变量,这里需要一个定位客户端类 LocationClient 和一个自定义的定位事件监听器 BDLocationListener 的实例,如下所示。

```
//定位相关
LocationClient mLocClient;
public MyLocationListenner myListener=
new MyLocationListenner();
//地图相关
MapView mMapView;
BaiduMap mBaiduMap;
//是否首次定位
boolean isFirstLoc=true;
public class MyLocationListenner implements BDLocationListener {
    @Override
    public void onReceiveLocation(BDLocation location) {
    }
}
```

BDLocationListener 用以监听实时从百度传回的定位信息,每次定位事件发生后都会调用 onReceiveLocation 方法。所以在该方法中写入定位信息,获取当前的经纬度和指定方向,如下所示。

```
//MapView 销毁后不再处理新接收的位置
if (location==null||mMapView==null)
    return;
MyLocationData locData=new MyLocationData.Builder()
    .accuracy(location.getRadius())
        //此处设置开发者获取到的方向信息,顺时针 0~360°
    .direction(0).latitude(location.getLatitude())
    .longitude(location.getLongitude()).build();
    mBaiduMap.setMyLocationData(locData);
//首次定位时更新地图状态
if (isFirstLoc) {
    isFirstLoc=false;
    LatLng ll=new LatLng(location.getLatitude(),
    location.getLongitude());
    MapStatusUpdate u=MapStatusUpdateFactory.newLatLng(ll);
    mBaiduMap.animateMapStatus(u);
}
```

其中,getLatitude()和 getLongitude()用以返回当前定位的精确经纬度,direction()方法中的参数是开发者传入的方向信息,范围为 0~360°,可以使用方向传感器获取。然后重写 onCreate()等生命周期中的方法。定位功能是一个新的图层,需要在开启定位图层并初始化,并在 onDestroy()方法中关闭并销毁,如下所示。

```
@Override
protected void onCreate(Bundle savedInstanceState) {
    super.onCreate(savedInstanceState);
    SDKInitializer.initialize(getApplicationContext());
    setContentView(R.layout.activity_map_demo);
    mMapView=(MapView) findViewById(R.id.bmapView);
    mBaiduMap=mMapView.getMap();
    //开启定位图层
    mBaiduMap.setMyLocationEnabled(true);
    mBaiduMap.setMyLocationConfigeration(new MyLocationConfiguration(
        LocationMode.NORMAL, true, null));
    //定位初始化
    mLocClient=new LocationClient(this);
    mLocClient.registerLocationListener(myListener);
    LocationClientOption option=new LocationClientOption();
    option.setOpenGps(true);               //打开 gps
    option.setCoorType("bd09ll");          //设置坐标类型
    option.setScanSpan(1000);
    mLocClient.setLocOption(option);
    mLocClient.start();
```

```
}
@Override
protected void onResume() {
    mMapView.onResume();
    super.onResume();
}
@Override
protected void onPause() {
    mMapView.onPause();
    super.onPause();
}
@Override
protected void onDestroy() {
    //退出时销毁定位
    mLocClient.stop();
    //关闭定位图层
    mBaiduMap.setMyLocationEnabled(false);
    mMapView.onDestroy();
    mMapView=null;
    super.onDestroy();
}
```

其中，使用 mBaiduMap.setMyLocationEnabled(true) 方法开启定位图层，使用 registerLocationListener() 方法将定位事件监听器注册在定位客户端中并配置参数，然后调用 start() 启动客户端接收。可以使用 mBaiduMap.setMyLocationConfigeration() 方法为定位图标设置样式和定位模式。这里有三种定位模式，即 NORMAL、FOLLOWING 和 COMPASS。默认为 NORMAL 模式，定位效果如图 13-15 所示。

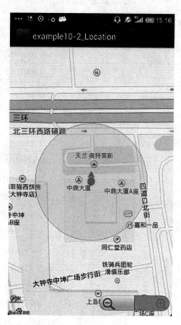

图 13-15 NORMAL 模式定位效果

可以布局一个 Button 用以选择定位模式，如下所示。

```
<Button
    android:id="@+id/button1"
    android:layout_width="wrap_content"
    android:layout_height="wrap_content"
    android:layout_alignParentRight="true"
    android:layout_alignParentTop="true"
    android:layout_marginRight="25dp"
    android:layout_marginTop="10dip"
    android:textColor="#000000" />
```

写入 Button 事件监听器，并执行导航模式切换，如下所示。

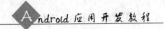

```
Button requestLocButton;
LocationMode mCurrentMode;
…
@Override
protected void onCreate(Bundle savedInstanceState) {
    super.onCreate(savedInstanceState);
    SDKInitializer.initialize(getApplicationContext());
    setContentView(R.layout.activity_map_demo);
    requestLocButton= (Button) findViewById(R.id.button1);
    mCurrentMode=LocationMode.NORMAL;
    requestLocButton.setText("普通");
    OnClickListener btnClickListener=new OnClickListener() {
        public void onClick(View v) {
            switch (mCurrentMode) {
                case NORMAL:
                    requestLocButton.setText("跟随");
                    mCurrentMode=LocationMode.FOLLOWING;
                    mBaiduMap.setMyLocationConfiguration(new
                        MyLocationConfiguration(mCurrentMode, true, null));
                    break;
                case COMPASS:
                    requestLocButton.setText("普通");
                    mCurrentMode=LocationMode.NORMAL;
                    mBaiduMap.setMyLocationConfiguration(new
                        MyLocationConfiguration(mCurrentMode, true,null ));
                    break;
                case FOLLOWING:
                    requestLocButton.setText("罗盘");
                    mCurrentMode=LocationMode.COMPASS;
                    mBaiduMap.setMyLocationConfiguration(new
                        MyLocationConfigeration(mCurrentMode, true,null ));
                    break;
            }
        }
    };
    requestLocButton.setOnClickListener(btnClickListener);
    …
```

使用 FOLLOWING 模式时，用户所在的定位地点将会被锁定在屏幕中心位置，不管怎样拖动地图都不会改变。使用 COMPASS 模式时，定位点位置会出现一个指北的罗盘，使用户能够明确方向，三种定位模式的比较如图 13-16 所示。

本案例以定位功能为例，带领读者大致了解了开发 LBS 应用中使用百度地图 Android SDK 的步骤及方法。如果读者想要了解并开发百度地图的其他相关功能，可以访问百度地图 API 官网，在本例的基础上，参阅 API 文档自行开发。

(a) NORMAL模式　　　　　　(b) FOLLOWING模式　　　　　　(c) COMPASS模式

图 13-16　三种定位模式效果

习 题 13

1. 什么是 LBS？LBS 有哪几种服务模式？
2. 简要说明 Android 的几种定位方式。
3. 说出几个现阶段 Android LBS 开发常用的第三方地图应用接口。

第 14 章

综合项目之手机监控

本章主要根据之前学习的知识,综合开发一款 Android 应用程序。该程序将带领读者复习和熟练使用之前介绍的 Android 组件、布局、UI 组件、动画和 Fragment 等相关内容。希望通过本章节的学习,使读者能更好地综合运用 Android 应用开发的相关知识,为今后从事基于 Android 的移动设备软件开发打下良好的基础。

本章将为读者介绍一个名为 Where Are You 的 Android 手机监控软件项目的研发。该软件的主要作用是帮助用户随时监控手机状态,或控制手机做一些操作,如自动拨打电话、开启警报音乐等。当用户手机不慎遗失时,能够通过另一部手机发送的相应指令,找到用户自身手机的位置。

14.1 项目功能需求分析

Android 手机监控软件项目需要实现的具体功能分析如下。

1. 拦截手机接收到的短信信息

当其他移动设备向该部手机发送短信时,软件可以获取该手机接收到的短信信息,获取的信息包括发送短信的手机号码及接收短信的具体内容。

2. 拦截手机接到的电话

当其他移动设备向该部手机拨打电话时,软件能够获得拨打电话人的手机号码。

3. 获取手机位置信息

当软件设置为手机自动开机运行时,能够随时获取当前所在的地理位置信息(经度和纬度)。

4. 手机自动回拨

通过另一个移动设备向装有该软件的手机发送指令,使手机能够自动地向指定手机拨打电话。

5. 手机自动播放铃音

通过另一个移动设备向装有该软件的手机发送指令,使手机能够自动地播放预先设定好的铃音。

项目最终预期要实现的软件主界面效果图如图 14-1 所示。

图 14-1　软件主界面效果图

14.2　应用程序结构设计

项目应用程序主要包含 4 类文件:布局文件、Activity 组件类、功能逻辑类和工具类文件。

1. 布局文件

包括三个布局文件:activity_main.xml、activity_welcome.xml、listview_item.xml 和 alpha_animation.xml,其功能如下。

(1) activity_welcome.xml:欢迎界面的布局文件;

(2) alpha_animation.xml:欢迎界面的动画布局文件;

(3) activity_main.xml:主功能界面的布局文件;

(4) listview_item.xml:主功能界面的 listview 列表项 Item 布局文件。

2. Activity 组件类

包含三个 Activity 组件类,即 MainActivity、WelcomeActivity 和 ListViewAdapter 类。以上类文件用于实现用户界面交互功能,其功能如下。

(1) WelcomeActivity 类：实现欢迎界面；

(2) MainActivity 类：主功能界面，调用各个功能逻辑类的方法，完成相关的业务逻辑；

(3) ListViewAdapter 类：继承 BaseAdapter，为界面 ListView 的适配器。

3. 功能逻辑类

(1) MyLocationListener 类：定位手机位置信息；

(2) SMSService 类：接收系统的短信广播。

4. 工具类

(1) ActionUtils 类：短信发送工具类；

(2) AppContext 类：全局的上下文类，封装共有信息。

14.3 应用程序界面设计

应用程序界面包括欢迎界面和主功能界面，其中欢迎界面顶层使用了 LinearLayout 布局，并在 LinearLayout 布局中添加了 ImageView 控件用来显示欢迎图片，并给图片添加了动画效果，提供用户体验。主功能界面比较复杂，它的最外层布局使用的是 RelativeLayout 布局，内部则是使用 TextView 控件显示标题，利用 EditText 控件接收用户输入，并采用 ListView 控件实现列表来展示所有的功能。

14.3.1 欢迎界面布局设计

欢迎界面布局文件为 activity_welcome.xml，代码如下。

```xml
<?xml version="1.0" encoding="utf-8"?>
<LinearLayout xmlns:android="http://schemas.android.com/apk/res/android"
    android:layout_width="match_parent"
    android:layout_height="match_parent"
    android:orientation="vertical">
    <ImageView
        android:id="@+id/iv_welcome_bg"
        android:layout_width="match_parent"
        android:layout_height="match_parent"
        android:background="@drawable/splash_bg"/>
</LinearLayout>
```

在欢迎界面布局中使用了 LinearLayout 作为整体布局，其宽和高匹配整个屏幕，在 LinearLayout 内部添加了一个 ImageView 视图控件用来显示一个 logo 图片。欢迎界面运行效果如图 14-2 所示。

图 14-2 欢迎界面运行效果

14.3.2 主功能界面布局设计

主功能界面布局文件为 activity_main.xml，代码如下。

```
<RelativeLayout
xmlns:android="http://schemas.android.com/apk/res/android"
xmlns:tools="http://schemas.android.com/tools"
android:id="@+id/RelativeLayout1"
android:layout_width="match_parent"
android:layout_height="match_parent"
android:background="@drawable/main_bei"
android:orientation="vertical"
android:paddingBottom="@dimen/activity_vertical_margin"
android:paddingLeft="@dimen/activity_horizontal_margin"
android:paddingRight="@dimen/activity_horizontal_margin"
android:paddingTop="@dimen/activity_vertical_margin"
tools:context=".MainActivity">

<LinearLayout
android:id="@+id/LinearLayout2"
android:layout_width="fill_parent"
android:layout_height="wrap_content"
android:layout_alignParentLeft="true"
android:layout_alignParentTop="true"
android:background="@drawable/title_bg"
android:gravity="center">
```

```xml
<ImageView
    android:id="@+id/iv_menu"
    android:layout_width="wrap_content"
    android:layout_height="wrap_content"
    android:layout_marginRight="60dp"
    android:src="@drawable/red_title_menu"/>

<TextView
    android:layout_width="wrap_content"
    android:layout_height="wrap_content"
    android:text="@string/title_content"
    android:textColor="#FFF"
    android:textSize="20sp"/>

<ImageView
    android:id="@+id/iv_change"
    android:layout_width="wrap_content"
    android:layout_height="wrap_content"
    android:layout_marginLeft="60dp"
    android:src="@drawable/red_title_change"/>
</LinearLayout>

<ListView
    android:id="@+id/lv1"
    android:layout_width="match_parent"
    android:layout_height="wrap_content"
    android:layout_alignParentLeft="true"
    android:layout_below="@+id/et_main_phonenumber"
    android:background="@android:color/white"
    android:layout_marginTop="10dp">
</ListView>

<EditText
    android:id="@+id/et_main_phonenumber"
    android:layout_width="match_parent"
    android:layout_height="wrap_content"
    android:layout_below="@+id/textView2"
    android:layout_marginTop="14dp"
    android:ems="10"
    android:inputType="phone"
    android:hint="@string/et_hint_content"
    android:textColor="@android:color/white"/>

<TextView
```

```xml
android:id="@+id/textView2"
android:layout_width="wrap_content"
android:layout_height="wrap_content"
android:layout_below="@+id/LinearLayout2"
android:layout_centerHorizontal="true"
android:layout_marginTop="15dp"
android:text="@string/et_hint_content"
android:textColor="@android:color/white"
android:textSize="15sp"/>

<Button
android:id="@+id/btn_start"
android:layout_width="130dp"
android:layout_height="wrap_content"
android:layout_alignParentBottom="true"
android:layout_alignParentLeft="true"
android:background="@drawable/server_start_p"
android:text="@string/btn_start_content"
android:textColor="@android:color/white"
android:textSize="20sp"/>

<Button
android:id="@+id/btn_stop"
android:layout_width="130dp"
android:layout_height="wrap_content"
android:layout_alignParentBottom="true"
android:layout_alignParentRight="true"
android:background="@drawable/server_off_p"
android:text="@string/btn_stop_content"
android:textColor="@android:color/white"
android:textSize="20sp"/>

</RelativeLayout>
```

在主功能界面布局中,使用了 RelativeLayout 作为整个界面的布局,其内部分成了4个部分进行布局设置,分别为"标题部分"、"主控手机输入"、"功能处理部分"和"监听控制部分"。

第一部分,使用了 LinearLayout 布局嵌套在 RelativeLayout 内部,并在其内设置了三个控件,分别是两个 ImageView 视图控件、一个 TextView 文本框控件。其中 TextView 控件是为了显示标题。

第二部分,在标题部分之下,主要包括 TextView 控件和 EditText 控件,其中 TextView 控件用来提示用户输入内容,而 EditText 控件是让用户输入主控手机号。当应用程序获取到消息时,可以通过用户输入的手机号,将消息发送到主控手机中。

第三部分，使用 ListView 控件，展示手机监控的主要功能，通过该列表用户就能够对应用程序进行操作。

第四部分，设置了两个 Button 按钮，分别用来进行监控的开启和关闭动作。

14.3.3 ListView 列表项 Item 布局

ListView 的列表项 Item 布局文件为 listview.xml，代码如下。

```xml
<?xml version="1.0" encoding="utf-8"?>
<RelativeLayout
    xmlns:android="http://schemas.android.com/apk/res/android"
    android:layout_width="match_parent"
    android:layout_height="match_parent">

    <ImageView
        android:id="@+id/listview_image"
        android:layout_width="wrap_content"
        android:layout_height="wrap_content"
        android:layout_alignParentLeft="true"
        android:layout_centerInParent="true"
        android:src="@drawable/iconlist_1"/>

    <LinearLayout
        android:layout_width="wrap_content"
        android:layout_height="wrap_content"
        android:layout_alignParentLeft="true"
        android:layout_centerInParent="true"
        android:layout_marginLeft="33dp"
        android:orientation="vertical">

        <TextView
            android:id="@+id/listview_bigtext"
            android:layout_width="wrap_content"
            android:layout_height="wrap_content"
            android:text="TextView1"
            android:textSize="15sp"/>

        <TextView
            android:id="@+id/listview_smalltext"
            android:layout_width="wrap_content"
            android:layout_height="wrap_content"
            android:text="TextView2"
            android:textSize="10sp"/>
    </LinearLayout>
```

```
<RadioButton
android:id="@+id/rb_item_state"
android:layout_width="wrap_content"
android:layout_height="wrap_content"
android:layout_alignParentRight="true"
android:layout_centerInParent="true"
android:layout_marginLeft="56dp"
android:button="@drawable/radio_button"/>

</RelativeLayout>
```

上述布局文件定义了 ListView 的显示格式，其顶层采用 RelativeLayout 布局，在其左边嵌入了 ImageView 控件显示列表项图标；又嵌入了一个 LinearLayout，垂直方向排列两个 TextView 控件，用来显示功能标题和功能说明；最右边嵌入了一个 RadioButton 控件作为功能开关按钮。

主功能界面的运行效果如图 14-3 所示。

图 14-3　主功能界面运行效果

14.4　Activity 类设计

14.4.1　欢迎界面 Activity

欢迎界面用来显示一个带有动画的 logo 图片，动画结束后会自动跳转到主功能界面。动画布局文件为 alpha_animation.xml，代码如下：

```xml
<?xml version="1.0"encoding="utf-8"?>
<alpha
xmlns:android="http://schemas.android.com/apk/res/android"
android:fromAlpha="0.5"
android:toAlpha="1"
android:duration="5000"
android:fillAfter="true"
android:interpolator="@android:anim/accelerate_decelerate_interpolator"
/>
```

在上述动画布局中,定义了透明度动画,其初始值为 0.5,表示欢迎界面中的图片是半透明显示,结束设置为 1 表示图片完全显示。这个动画过程持续执行 5000ms,通过属性设置 android:fillAfter="true"将图片保持在动画结束的位置上,同时也通过属性 android:interpolator 让图片先快后慢地显示出来。

欢迎界面 Activity 的类文件为 WelcomeActiviy.java,其代码如下。

```java
package com.example.android_demo14_1;

import android.app.Activity;
import android.content.Intent;
import android.os.Bundle;
import android.view.Window;
import android.view.animation.Animation;
import android.view.animation.Animation.AnimationListener;
import android.view.animation.AnimationUtils;
import android.widget.ImageView;

public class WelcomeActivity extends Activity implements AnimationListener {
    /**
     * 视图控件
     */
    private ImageView welcomeIv;
    /**
     * 动画成员
     */
    private Animation anim;

    /**
     * 初始化方法
     */
    private void setupView() {
        welcomeIv= (ImageView) this.findViewById(R.id.iv_welcome_bg);
        anim=AnimationUtils.loadAnimation(this, R.anim.alpha_animation);
    }
```

```java
/**
 * 添加监听器方法
 */
private void addListener() {
    anim.setAnimationListener(this);
}

@Override
protected void onCreate(Bundle savedInstanceState) {
    super.onCreate(savedInstanceState);
    //去掉标题
    requestWindowFeature(Window.FEATURE_NO_TITLE);
    //设置布局文件
    setContentView(R.layout.activity_welcome);
    //引用初始化方法
    setupView();
    //引用添加监听器方法
    addListener();
    //开启动画
    welcomeIv.startAnimation(anim);
}

@Override
public void onAnimationStart(Animation animation) {
    //TODO Auto-generated method stub
}

@Override
public void onAnimationEnd(Animation animation) {
    //动画结束,跳转到主界面
    //声明 Intent 对象,并指定目标
    Intent intent=new Intent(this, MainActivity.class);
    //开启 Activity
    startActivity(intent);
    //关闭 Activity
    finish();
}

@Override
public void onAnimationRepeat(Animation animation) {
    //TODO Auto-generated method stub
}
}
```

在 WelcomeActivity 类的 onCreate() 方法中引用了 setupView() 方法,该方法通过

findViewById()方法获得视图控件,并在其中通过 AnimationUtile 类的 loadAnimation()方法将 res/anim 中的动画布局文件 alpha_animation.xml 加载到 Animation 类中,同时返回 Animation 对象。

动画在开始之前,要为动画添加 AnimationListener 监听器,目的是为了当动画结束时,能够捕捉到 Animation 的结束状态,进行界面跳转,并调用 Context.finish()方法关闭欢迎界面。

14.4.2 主功能界面 Activity

主功能界面 Activity 的类文件为 MainActivity.java,其代码如下。

```
package com.example.android_demo14_1;

import android.app.Activity;
import android.content.Intent;
import android.os.Bundle;
import android.view.View;
import android.view.View.OnClickListener;
import android.view.Window;
import android.widget.Button;
import android.widget.EditText;
import android.widget.ListView;
import android.widget.Toast;

public class MainActivity extends Activity implements OnClickListener {
    /**
     * 数据资源
     */
    private int[] images={ R.drawable.iconlist_1, R.drawable.iconlist_2,
            R.drawable.iconlist_3, R.drawable.iconlist_4,
            R.drawable.iconlist_5, R.drawable.iconlist_6 };
    private String[] bigText={ "短信监控","来电监控","位置监控","回电监控","控制铃声" };
    private String[] smallText={ "将接收到短信内容及收件人转发到主控手机","将呼入电话号码发送到主控手机","将被监控手机当前位置发送到主控手机","控制被控手机拨打电话到主控手机","控制被控手机铃声" };
    /**
     * 功能标记数组
     */
    private boolean[] isAction=new boolean[5];
    /**
     * 列表控件
     */
    private ListView listview;
```

```java
/**
 * 输入框控件
 */
private EditText phoneNumberEt;
/**
 * 适配器控件
 */
private ListViewAdapter adapter;
/**
 * 开启监听和关闭监听按钮
 */
private Button startBtn, stopBtn;
/**
 * 监听意图
 */
private Intent stateIntent;

/**
 * 初始化控件对象
 */
private void setupView() {
    listview=(ListView) this.findViewById(R.id.lv1);
    phoneNumberEt=(EditText) this.findViewById(R.id.et_main_phonenumber);
    startBtn= (Button) this.findViewById(R.id.btn_start);
    stopBtn= (Button) this.findViewById(R.id.btn_stop);
}

/**
 * 初始化 Adapter 对象
 */
private void initAdapter() {
    adapter=new ListViewAdapter(images, bigText, smallText, isAction,
            this);
}

/**
 * 添加监听事件
 */
private void addListener() {
    startBtn.setOnClickListener(this);
    stopBtn.setOnClickListener(this);
}

@Override
```

```java
protected void onCreate(Bundle savedInstanceState) {
    super.onCreate(savedInstanceState);
    //去掉标题
    requestWindowFeature(Window.FEATURE_NO_TITLE);
    //设置布局文件
    setContentView(R.layout.activity_main);
    //引用初始化方法
    setupView();
    //初始化适配器方法
    initAdapter();
    //引用添加监听器方法
    addListener();
    //缓存功能状态
    AppContext.setIsAction(isAction);
    //初始化意图对象
    stateIntent=new Intent(this, SMSService.class);
    //ListView 设置适配器
    listview.setAdapter(adapter);
}

@Override
public void onClick(View v) {
    int id=v.getId();
    switch (id) {
    case R.id.btn_start:              //开启监听器
        //获取输入框中的主控手机号
        String phoneNumber=phoneNumberEt.getText().toString().trim();
        //判断主控手机是否为空
        if ("".equals(phoneNumber)) {
            //为空,提示错误信息
            phoneNumberEt.setError(getResources().getString(
                R.string.toast_phone_number_null));
            return;
        }
        //判断当前监听状态
        if (AppContext.isStates()) {
            //提示不能重复单击
            Toast.makeText(this,R.string.toast_no_stop,Toast.LENGTH_LONG)
                .show();
            return;
        }
        //修改监听状态
        AppContext.setStates();
        //将主控手机号码存储到缓存类中
```

```
            AppContext.setPhoneNumber(phoneNumber);
            //设置输入框不能输入
            phoneNumberEt.setEnabled(false);
            //开启服务
            startService(stateIntent);
            break;
        case R.id.btn_stop:                        //关闭监听器
            //判断当前监听状态
            if (!AppContext.isStates()) {
                //提示不能重复单击
                Toast.makeText(this,R.string.toast_no_start,Toast.LENGTH_LONG)
                        .show();
                return;
            }
            //修改监听状态
            AppContext.setStates();
            //设置输入框能输入
            phoneNumberEt.setEnabled(true);
            //停止服务
            stopService(stateIntent);
            break;
        default:
            break;
        }
    }
}
```

上述代码中，主要实现了监听服务的开启和关闭，并通过自定义的 ListViewAdapter 实现了数据的加载。同时还为开启和关闭监听按钮添加了单击事件，在单击事件中判断按钮是否被重复启动，如果重复启动将不予执行。需要注意，在开启监听之前，要保证用户已经输入了主控手机号，如果没有输入，将提示错误信息。主控手机号没有输入，单击开启监听服务按钮时的效果图如图 14-4 所示。

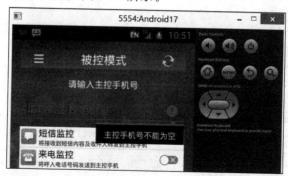

图 14-4　主控手机号为空时开启监听服务的错误提示

在 MainActivity 类中用到了自定义的 ListViewAdapter 类，文件为 ListViewAdapter.java，代码如下。

```java
package com.example.android_demo14_1;

import android.content.Context;
import android.view.LayoutInflater;
import android.view.View;
import android.view.View.OnClickListener;
import android.view.ViewGroup;
import android.widget.BaseAdapter;
import android.widget.ImageView;
import android.widget.RadioButton;
import android.widget.TextView;
/**
 * ListView 适配器
 */
public class ListViewAdapter extends BaseAdapter{
    /**
     * 数据
     */
    private int[] images;
    private String[] bigText;
    private String[] smallText;
    private boolean[] isAction;
    //上下文对象
    private Context context;
    //布局加载器
    private LayoutInflater inflater;

    /**
     * 构造方法
     * @param images
     * @param bigText
     * @param smallText
     * @param isAction
     * @param context
     */
    public ListViewAdapter(int[] images, String[] bigText, String[] smallText,
            boolean[] isAction, Context context) {
        super();
        this.images=images;
        this.bigText=bigText;
        this.smallText=smallText;
```

```java
        this.isAction=isAction;
        this.context=context;
        this.inflater=LayoutInflater.from(context);
    }

    @Override
    public int getCount() {
        if(bigText !=null){
            return bigText.length;
        }
        return 0;
    }

    @Override
    public Object getItem(int position) {
        //TODO Auto-generated method stub
        return null;
    }

    @Override
    public long getItemId(int position) {
        //TODO Auto-generated method stub
        return 0;
    }

    @Override
    public View getView(final int position, View convertView, ViewGroup parent)
    {
        //获得布局控件
        View view=inflater.inflate(R.layout.listview_item,
                null);
        //获得布局中的控件
        TextView bigtext= (TextView) view
                .findViewById(R.id.listview_bigtext);
        TextView smalltext= (TextView) view
                .findViewById(R.id.listview_smalltext);
        ImageView listview_image= (ImageView) view
                .findViewById(R.id.listview_image);
        final RadioButton actionRb= (RadioButton) view.findViewById(R.id.rb_item_state);
        //设置单选按钮状态
        actionRb.setChecked(isAction[position]);
        //设置单选按钮单击事件
        actionRb.setOnClickListener(new OnClickListener() {
```

```java
            public void onClick(View v) {
                //设置单选按钮状态
                actionRb.setChecked(!isAction[position]);
                //修改缓存状态
                isAction[position]=!isAction[position];
                //修改动作数组状态
                AppContext.setIsAction(isAction);
            }
        });
        //显示文本
        if(bigText !=null){
            //设置数据
            bigtext.setText(bigText[position]);
            smalltext.setText(smallText[position]);
            listview_image.setImageResource(images[position]);
        }
        //返回控件
        return view;
    }
}
```

上述代码中,通过继承 BaseAdapter 类实现自定义的 ListViewAdapter 适配器,并在适配器中的 getView()方法中获得 View 布局控件,然后通过 LayoutInflater.inflate()方法,根据 View 控件获取 Item 布局里面的所有控件。需要注意,在设置 RadioButton 状态时用到了 setChecked()方法,true 代表选中,false 代表未选中。

14.5 应用程序主要功能逻辑设计

应用程序界面和 Activity 类设计开发完毕之后,就可以为程序添加主要的处理逻辑。当前的项目中主要的功能包括拦截短信/电话、定位、回拨电话和控制铃音。电话拦截使用的是广播技术,通过接收系统发送的短信广播就可以获得短信中的详细信息;而电话拦截用到了 TelephonyManager 管理器,通过添加 PhoneStateListener 接口可以获得当前电话的状态。手机定位使用了百度定位 SDK(详细信息请参考百度开发平台 http://developer.baidu.com/map/index.php?title=android-locsdk)。

14.5.1 服务类 SMSService

SMSService 类主要用于接收系统的短信广播,实现短信/电话的拦截,并调用动作工具类 ActionUtils 进行相应的动作;另外一个功能是初始化百度定位 SDK,开启定位服务。SMSService 类是应用程序的主要逻辑类,其类文件为 SMSService.java,代码如下:

```java
package com.example.android_demo14_1;

import java.text.SimpleDateFormat;
import java.util.Date;

import android.app.Service;
import android.content.BroadcastReceiver;
import android.content.Context;
import android.content.Intent;
import android.content.IntentFilter;
import android.os.Bundle;
import android.os.IBinder;
import android.telephony.PhoneStateListener;
import android.telephony.SmsMessage;
import android.telephony.TelephonyManager;
import android.widget.Toast;

import com.baidu.location.BDLocationListener;
import com.baidu.location.LocationClient;
import com.baidu.location.LocationClientOption;

/**
 *服务类
 *
 */
public class SMSService extends Service {
    //广播接收者
    private BroadcastReceiver receiver;
    //接口对象
    public LocationClient mLocationClient=null;
    BDLocationListener myListener=new MyLocationListener();

    @Override
    public IBinder onBind(Intent intent) {
        //TODO Auto-generated method stub
        return null;
    }

    @Override
    public void onCreate() {
        super.onCreate();
        //声明LocationClient类
        mLocationClient=new LocationClient(getApplicationContext());
        mLocationClient.registerLocationListener(myListener);    //注册监听函数
```

```java
        LocationClientOption option=new LocationClientOption();
        option.setOpenGps(true);
        option.setAddrType("all");       //返回的定位结果包含地址信息
        option.setCoorType("bd09ll");    //返回的定位结果是百度经纬度,默认值为gcj02
        option.setScanSpan(5000);        //设置发起定位请求的间隔时间为5000ms
        option.disableCache(true);       //禁止启用缓存定位
        option.setPoiNumber(5);          //最多返回POI个数
        option.setPoiDistance(1000);     //POI查询距离
        option.setPoiExtraInfo(true);    //是否需要POI的电话和地址等详细信息
        mLocationClient.setLocOption(option);
        //开启定位
        mLocationClient.start();
    }

    @Override
    public int onStartCommand(Intent intent, int flags, int startId) {
        Toast.makeText(SMSService.this, "开启监听方法",
            Toast.LENGTH_LONG).show();
        //判断是否开启短信拦截
        if (AppContext.getIsAction()[0]) {
            //获得短信
            receiver=new BroadcastReceiver() {
                @Override
                public void onReceive(Context context, Intent intent) {
                    Bundle bundle=intent.getExtras();
                    if (bundle !=null) {
                        //获得原始数据
                        Object[] object=(Object[]) bundle.get("pdus");
                        SmsMessage[] messages=new SmsMessage[object.length];
                        for (int i=0; i<messages.length; i++) {
                            messages[i]=SmsMessage
                                .createFromPdu((byte[]) object[i]);
                        }
                        for (SmsMessage smsMessage : messages) {
                            //短信的内容
                            String body=smsMessage.getDisplayMessageBody();
                            //发送短信的电话号码
                            String address=smsMessage
                                .getDisplayOriginatingAddress();

                            //判断是否为主控手机的手机号
                            if (address.equals(AppContext.getPhoneNumber())) {
                                if ("callback".equals(body)
                                    && AppContext.getIsAction()[3]) {
```

```java
                    //回拨动作
                    ActionUtils.callback(
                        AppContext.getPhoneNumber(),
                        context);
                }
                if ("ring".endsWith(body)
                    && AppContext.getIsAction()[4]) {
                    //报警
                    ActionUtils.ring(context);
                }
                if ("location".endsWith(body)
                    && AppContext.getIsAction()[2]) {
                    if (mLocationClient !=null
                        && !mLocationClient.isStarted()) {
                        mLocationClient.start();
                    }
                    //发送短信
                    ActionUtils.sendSMS(
                        AppContext.getPhoneNumber(),
                        "latitude:"+AppContext.latitude
                        +",longitude:"
                        +AppContext.longitude);
                }
                if ("unlocation".equals(body)
                    && AppContext.getIsAction()[2]) {
                    if(mLocationClient !=null
                        && mLocationClient.isStarted()) {
                        mLocationClient.stop();
                    }
                    //发送短信
                    ActionUtils.sendSMS(
                        AppContext.getPhoneNumber(),
                        "Location Stop");
                }
            } else {
                StringBuffer sb=new StringBuffer();
                sb.append("Address:"+address);
                sb.append("Body:"+body);
                //发送短信
                ActionUtils.sendSMS(
                    AppContext.getPhoneNumber(),
                    sb.toString());
            }
        }
```

```java
                }
            }
        };
        //声明一个过滤器
        IntentFilter filter=new IntentFilter();
        //动作
        filter.addAction("android.provider.Telephony.SMS_RECEIVED");
        //设置过滤器的级别
        filter.setPriority(Integer.MAX_VALUE);
        //注册过滤器
        registerReceiver(receiver, filter);
    }
    //判断是否开启电话拦截
    if (AppContext.getIsAction()[1]) {
        //实现一个电话拦截
        TelephonyManager manager=(TelephonyManager) this
                .getSystemService(TELEPHONY_SERVICE);
        manager.listen(new PhoneStateListener() {
            @Override
            public void onCallStateChanged(int state, String incomingNumber) {
                super.onCallStateChanged(state, incomingNumber);
                //判断电话铃声响起
                if (state==TelephonyManager.CALL_STATE_RINGING) {
                    Date date=new Date();
                    SimpleDateFormat format=new SimpleDateFormat(
                            "yyyy:MM:dd");
                    //发送短信
                    ActionUtils.sendSMS(AppContext.getPhoneNumber(),
                            "PhoneNumber:"+incomingNumber+",Data:"
                                    +format.format(date));
                }
            }
        }, PhoneStateListener.LISTEN_CALL_STATE);
    }
    return super.onStartCommand(intent, flags, startId);
}

@Override
public void onDestroy() {
    super.onDestroy();
    if (receiver !=null) {
        unregisterReceiver(receiver);
    }
    mLocationClient.stop();
```

           }
       }
    当服务被开启的时候,首先会判断缓存类的功能数组中的对应功能是否开启。如当短信拦截的功能被开启时,服务中的 BroadcastReceiver 广播接收者会拦截系统发送的短信广播,并解析短信中的内容和电话号码;若服务判断当前短信不是主控手机发送的,就将发送过来的短信转发到主控手机上。

    在图 14-5 所示的界面中,我们向 5554 发送了一条短信,发送短信的号码为 1234566,内容为 welcome android。当被控模拟器收到短信时,发现不是主控手机的号码,于是就将短信内容转发到了主控手机(5556 模拟器)上,5556 模拟器就收到了一条来自被控手机的短信。

图 14-5　短信拦截运行效果

    而获得来电控制就是当被控手机有电话打入的时候,程序就会通过 PhoneStateListener 接口获得到手机状态,并将该电话的号码和拨打时间编辑成短信,发送到主控手机上。电话拦截运行效果如图 14-6 所示。

    当被控手机接收到主控手机的短信时,程序服务会判断是否为预定好的命令,例如我们设定的回拨命令为 callback,而接收到的短信也是主控手机发来的 callback 命令,被控手机会自动向主控手机拨打电话。回拨电话运行效果如图 14-7 所示。

    在 SMSService 类的 onCreate()方法中,对百度定位 SDK 的地图定位 API 进行了初始化并开启了定位服务。百度定位 SDK 需要到百度开发平台上下载,然后将其中的 liblocSDK3.so 文件拷贝到项目工程中的 libs/armeabi 目录下,并在工程中添加 JAR 包 locSDK3.3.jar。

    在百度定位 SDK 中,提供了一个定位服务,该服务需要在 AndroidManifest.xml 文件的 application 标签中声明,为了避免各个应用程序共用一个服务出现的权限问题,百度新版本定位 SDK 可以让应用程序单独拥有自己的定位服务,其参考代码片段如下:

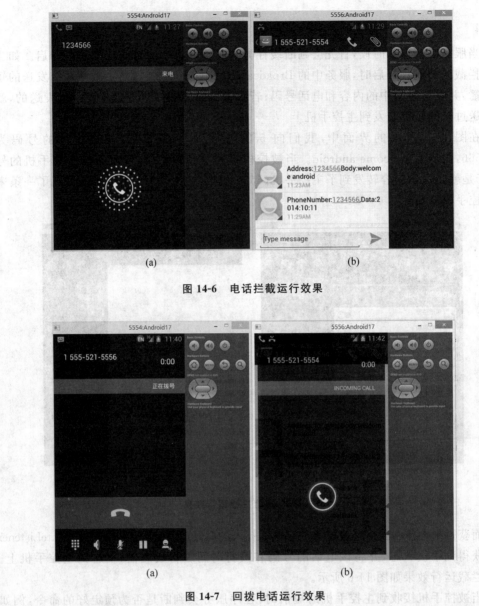

图 14-6 电话拦截运行效果

图 14-7 回拨电话运行效果

```
<service
android:name="com.baidu.location.f"
android:enabled="true"
android:process=":remote">
</service>
```

百度定位 SDK 在使用前同样也需要在 AndroidManifest.xml 文件中添加相应的权限,参考代码片段如下。

```
<uses-permission android:name="android.permission.ACCESS_COARSE_LOCATION">
```

```xml
</uses-permission>
<uses-permission android:name="android.permission.ACCESS_FINE_LOCATION">
</uses-permission>
<uses-permission android:name="android.permission.ACCESS_WIFI_STATE">
</uses-permission>
<uses-permission android:name="android.permission.ACCESS_NETWORK_STATE">
</uses-permission>
<uses-permission android:name="android.permission.CHANGE_WIFI_STATE">
</uses-permission>
<uses-permission android:name="android.permission.READ_PHONE_STATE">
</uses-permission>
<uses-permission android:name="android.permission.WRITE_EXTERNAL_STORAGE">
</uses-permission>
<uses-permission
android:name="android.permission.INTERNET"/>
<uses-permission
android:name="android.permission.MOUNT_UNMOUNT_FILESYSTEMS">
</uses-permission>
<uses-permission
android:name="android.permission.READ_LOGS">
</uses-permission>
```

在 SMSService 类的 onCreate() 方法中主要是初始化 LocationClient 类，此处需要注意，LocationClient 类必须在主线程中声明，且需要 Context 类型的参数。Context 需要是全进程有效的 context，推荐用 getApplicationConext 获取全进程有效的 context。

14.5.2　获取定位信息类 MyLocationListener

MyLocationListener 类通过百度定位 SDK 的 BDLocationListener 接口，可以获取手机的位置信息，MyLocationListener 类文件为 MyLocationListener.java，代码如下。

```java
package com.example.android_demo14_1;

import com.baidu.location.BDLocation;
import com.baidu.location.BDLocationListener;
/**
 *定位接口实现类
 */
public class MyLocationListener implements BDLocationListener {

    @Override
    public void onReceiveLocation(BDLocation location) {
        if (location==null)
            return ;
        AppContext.latitude=location.getLatitude();
```

```
        AppContext.longitude=location.getLongitude();

    }

    @Override
    public void onReceivePoi(BDLocation arg0) {
        //TODO Auto-generated method stub

    }
}
```

上述代码中,通过实现 BDLocationListener 接口来实现其内部的两个方法,这两个方法的主要作用如下:

(1) 接收异步返回的定位结果,参数是 BDLocation 类型参数;

(2) 接收异步返回的 POI 查询结果,参数是 BDLocation 类型参数。

定位效果如图 14-8 所示。

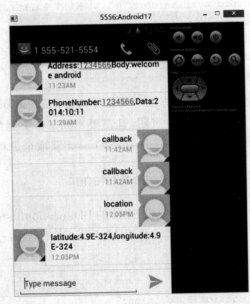

图 14-8　获得定位信息

14.6　工具类设计

14.6.1　缓存类 AppContext

AppContext 类是全局的上下文类,封装共有信息,存储程序中的相关数据,类文件为 AppContext.java,代码如下。

```
package com.example.android_demo14_1;
```

```java
/**
 * 缓存类
 */
public class AppContext {
    /**
     * 保存电话号码
     */
    private static String phoneNumber;
    /**
     * 保存功能状态
     */
    private static boolean[] isAction;
    /**
     * 监听状态
     */
    private static boolean isStates;
    /**
     * 纬度
     */
    public static double latitude ;
    /**
     * 经度
     */
    public static double longitude ;
    public static boolean isStates() {
        return isStates;
    }
    public static void setStates() {
        AppContext.isStates=!isStates;
    }
    public static String getPhoneNumber() {
        return phoneNumber;
    }
    public static void setPhoneNumber(String phoneNumber) {
        AppContext.phoneNumber=phoneNumber;
    }
    public static boolean[] getIsAction() {
        return isAction;
    }
    public static void setIsAction(boolean[] isAction) {
        AppContext.isAction=isAction;
    }
}
```

14.6.2 动作工具类 ActionUtils

ActionUtils 类将一些经常使用的动作进行了封装，这样可以更好地简化程序的复杂度，主要定义了三个动作方法，分别是"发送短信"、"回拨电话"和"播放铃音"。ActionUtils 类文件为 ActionUtils.java，代码如下。

```java
package com.example.android_demo14_1;

import android.content.Context;
import android.content.Intent;
import android.media.MediaPlayer;
import android.net.Uri;
import android.telephony.SmsManager;

public class ActionUtils {

    /**
     * 发送短信类
     * SEND_SMS
     * @param phone
     * @param content
     */
    public static void sendSMS(String phone, String content){
        SmsManager manager=SmsManager.getDefault();
        manager.sendTextMessage(phone, null, content, null, null);
    }

    /**
     * 回拨电话
     * CALL_PHONE
     * @param phone
     * @param context
     */
    public static void callback(String phone,Context context){
        Intent intent=new Intent();
        intent.setAction(Intent.ACTION_CALL);
        intent.setData(Uri.parse("tel:"+phone));
        intent.setFlags(Intent.FLAG_ACTIVITY_NEW_TASK);
        context.startActivity(intent);
    }
    /**
     * 播放铃音
     * @param context
     */
    public static void ring(Context context){
        MediaPlayer player=MediaPlayer.create(context, R.raw.prisonbreak);
        player.start();
    }
}
```

习 题 14

1. 在本章的综合项目中,欢迎界面的动画是如何实现的?请简要说明。

2. 在 Android 的应用程序中,如何获取当前来电的电话号码?如何获取接收到短信的发送方电话号码及短信内容?

3. 在本章的综合项目中,为了实现应用程序访问网络、GPS 定位等功能,需要在 AndroidManifest.xml 文件中添加什么权限?请具体说明。

参 考 文 献

[1] http://developer.android.com/guide
[2] http://developer.android.com/reference
[3] http://developer.android.com/samples
[4] http://android-developers.blogspot.com
[5] https://www.google.com/design/spec/components/bottom-sheets.html#bottom-sheets-specs
[6] https://plus.google.com/+AndroidDevelopers/posts
[7] http://stackoverflow.com/questions/tagged/android
[8] https://sites.google.com/a/android.com/tools/recent
[9] Marko Gargenta,Masumi Nakamura. Learning Android 中文版. 卢涛,李颖译. 第 2 版. 北京：电子工业出版社,2014.
[10] Paul Deitel,Harvey Deitel,Abbey Deitel. Android 大学教程. 胡彦平,张君施,闫锋欣,等译. 第 2 版. 北京：电子工业出版社,2015.
[11] 任玉刚. Android 开发艺术探索. 北京：电子工业出版社,2015.
[12] 李刚. 疯狂 Android 讲义. 第 2 版. 北京：电子工业出版社,2015.
[13] Joseph Annuzzi Jr.,Lauren Darcey,Shane Conder. Android 应用程序开发权威指南. 林学森,周昊来,等译. 第 4 版. 北京：电子工业出版社,2015.
[14] 明日科技. Android 从入门到精通. 北京：清华大学出版社,2012.
[15] Brian Hardy,Bill Phillips. Android 编程权威指南. 王明发译. 北京：电子工业出版社,2014.